78th Annual Meeting of the Northeastern Weed Science Society 2024

Proceedings – Volume 78

Boston, Massachusetts, USA
8 - 11 January 2024

ISBN: 979-8-3313-0918-3

Printed from e-media with permission by:

Curran Associates, Inc.
57 Morehouse Lane
Red Hook, NY 12571

Some format issues inherent in the e-media version may also appear in this print version.

Northeastern Weed Science Society (NEWSS)
P.O. Box 25
Woodstown, NJ 08098
USA

Phone: 315-209-7580

northeasternweedscience@hotmail.com

Additional copies of this publication are available from:

Curran Associates, Inc.
57 Morehouse Lane
Red Hook, NY 12571 USA
Phone: 845-758-0400
Fax: 845-758-2633
Email: curran@proceedings.com
Web: www.proceedings.com

PROCEEDINGS - VOLUME 78

78th Annual Meeting of the Northeastern Weed Science Society (NEWSS)

Site of 2024 annual meeting:

January 8-11, 2024

HYATT REGENCY

1 Avenue de Lafayette, Boston, MA

§ Cover: The winning image in the 2024 NEWSS Photo Contest, submitted by Navdeep Godara, Virginia Tech

Proceedings
of the
Seventy Eighth Annual Meeting
of the
Northeastern Weed Science Society

Vijay Singh, Editor

NORTHEASTERN WEED SCIENCE SOCIETY

2024 SUSTAINING MEMBERS

PLATINUM

BASF We create chemistry

Bayer CropScience

CORTEVA agriscience

syngenta

SILVER

AMVAC An American Vanguard Company

FMC

pbi/gordon companies Employee-Owned

BRONZE

GDM

LABServices The life-science technology support team

Nufarm

Sustaining members: https://www.newss.org/about/sustaining-members/

Table of Contents

NEWSS EXECUTIVE COMMITTEE – OFFICERS

President	**Wes Everman** North Carolina State University wes_everman@ncsu.edu
President-elect	**Erin Hitchner** Syngenta erin.hitchner@syngenta.com
Vice President	**Thierry Besancon** Rutgers University thierry.besancon@rutgers.edu
Treasurer	**Larissa Smith** Syngenta larissa.smith@syngenta.com
Past President	**Jacob Barney** Virginia Tech jnbarney@vt.edu

NEWSS COMMITTEES – CHAIRS

Editor	Vijay Singh	Virginia Tech
Membership Chair	Matt Elmore	Rutgers University
Public Relations	John Kaminski	Penn State University
Research and Education	Carolyn Lowry	Penn State University
Sustaining Membership	John Wallace	Penn State University
CAST Representative	Anthony Witcher	Tennessee State University
Graduate Student Rep	Eric Jones	North Carolina State University
WSSA Representative	Steve Pyle	Syngenta
Science Policy Director	Lee vanWychen	WSSA

NEWSS COMMITTEES – CHAIRS

Student Paper Contest	Art Gover	Penn State University
Student Poster Contest	Anthony Witcher	Tennessee State University
Student Paper Writing Contest	Art Gover	Penn State University
Photo Judging Chair	Hilary Sandler	University of Massachusetts
Site Selection Committee	Wesley Everman	North Carolina State University
Awards Committee	Caren Schmidt	BASF
Nomination Committee	Matt Elmore	Rutgers University
Resolutions Committee	Daniel Kunkel	AMVAC
Collegiate Weed Contest Committee	Wesley Everman	North Carolina State University
Audit Committee	Larissa Smith	Syngenta
Archives Committee	Art Gover	Penn State University
Past Presidents Committee	Caren Schmidt	BASF
Endowment Committee	Caren Schmidt	BASF

2024 NEWSS PROGRAM

Program Committee	Erin Hitchner (Chair)
	Thierry Besancon
	Katherine Diehl
	Eli Russell
Section Chairs	
Agronomy	Vijay Singh
Weed Ecology and Invasive Species	Jacob barney
Turf, Ornamental, and Non-Crop	Matthew Elmore
Research Posters	Anthony L. Witcher
Fruits, Vegetables, and Tree Crops	Kurt Vollmer

2024 NEWSS AWARDS

NEWSS Fellow Award
Randy Prostak, University of Massachusetts

NEWSS Award of Merit
Hilary Sandler, University of Massachusetts

NEWSS Outstanding Researcher Award
Vijay Singh, Virginia Tech

Garry Schnappinger Service Recognition Award
Sudeep Mathew, Syngenta

Robert D. Sweet Outstanding Graduate Student – Ph.D.
Uriel Menalled, Cornell University

2024 NEWSS CONTEST WINNERS

Poster Winners

First Place	Navdeep Godara, Virginia Tech
Second Place	Daewon Koo, Virginia Tech

Oral Competition Winners

First Place	Aleah Butler-Jones, Cornell University
Second Place	Eli Russell, Virginia Tech

Graduate Student Writing Contest

First Place	Laurel Wellman, Penn State University
Second Place	Navdeep Godara, Virginia Tech
Second Place	Noelle Connors, Penn State University

Photo Contest

First Place	Navdeep Godara, Virginia Tech
Second Place	Juan Romero Cubas, Virginia Tech
Third Place	Sohaib Chatta, Clemson University

RESPONSES OF WEEDS AND MICROARTHROPOD COMMUNITIES TO ZASSO ELECTRICAL WEEDING.

Aleah L. Butler-Jones*[1], Elizabeth C. Maloney[2], John Owens[1], Gregory M. Peck[1], Kyle Wickings[2], Marcelo L. Moretti[3], Brad Hanson[4], Lynn M. Sosnoskie[2]; [1]Cornell University, Ithaca, NY, [2]Cornell University, Geneva, NY, [3]Oregon State University, Corvallis, OR, [4]University of California, Davis, Davis, CA (1)

The continued evolution of herbicide resistant weeds and growing consumer demand for non-chemical-based weed management have sparked growers' interest in novel technologies, including electrical weed control (EWC). Despite growing interest in EWC, there is limited information available regarding its impact on weed and soil communities. The objective of this research is to evaluate the impact of electrical weeding on weed cover, weed biomass, and soil microarthropod communities. To accomplish these goals, a tractor-mounted and PTO-driven Zasso Electroherb unit was operated in weedy, unplanted agricultural fields at two travel speeds (1.9 km/h, 4.8 km/h) and three amperage settings (mean 9.1 Amps, mean 19.5 Amps, mean 35.5 Amps). The experiment was organized as a randomized complete block design (RCBD) with four replicates and two experimental runs. Dominant weed species across sites included common lambsquarters (*Chenopodium album* L.), Powell amaranth (*Amaranthus powellii* S. Wats.), common ragweed (*Ambrosia artemisiifolia* L.), large crabgrass (*Digitaria sanguinalis* (L.) Scop.), yellow foxtail (*Setaria pumila* (Poir.) Roem. & Schult.), and yellow nutsedge (*Cyperus esculentus* L.). Estimates of weed cover (assessed using Canopeo) were collected within 24 hours of treatment and up to 20 days after treatment (20 DAT). Weed biomass was collected 28 DAT. The possible impacts of electrical current on microarthropod numbers will be determined using soil samples collected from the trials. Electrical weeding significantly reduced weed cover 20 DAT and weed biomass 28 DAT compared to the untreated check ($P < 0.05$); however, percent weed cover and biomass were not significantly different between the EWC treatments. Preliminary results from 2022 suggest that EWC does not reduce the total number of microarthropods, mites, or collembola ($P > 0.05$); however, 2023 data is currently being processed. In addition to this experiment, electrical weeding was compared to cultivation and organic herbicide application in established orchards. Preliminary results suggest that EWC is as if not more effective than cultivation at reducing weed cover and biomass. Further, EWC followed by cultivation more effectively controlled weeds compared to the sole use of either tactic. These trials will be expanded in 2024.

EFFICACY AND SAFETY OF PYRIDATE IN MID-ATLANTIC SWEET CORN.

Lynn M. Sosnoskie*[1], Thierry E. Besancon[2], Mark J. VanGessel[3], Dwight Lingenfelter[4]; [1]Cornell University, Geneva, NY, [2]Rutgers University, Chatsworth, NJ, [3]University of Delaware, Georgetown, DE, [4]Penn State University, University Park, PA (2)

Tank mixing 4-hydroxyphenylpyruvate dioxygenase (HPPD) inhibiting herbicides with triazine photosystem II (PS-II) inhibiting herbicides is recommended for POST applications in corn because of the synergistic response. The use of atrazine is impacted by the evolution of herbicide resistance and growing concerns regarding proposed US EPA restrictions on use rates and supplemental mitigation requirements. Pyridate is also a PS II-inhibitor that could be an effective replacement option to triazines. In 2023, research trials were conducted in NY, NJ, and PA to describe the impacts of tank-mixing pyridate with HPPD-inhibiting herbicides on weed control and crop safety in sweet corn ('Bodacious' [NY], 'Milky Way' [NJ], and 'Serendipity' [PA]). The main herbicide treatments included pyridate at 0.35 and 0.71 kg ha-1, atrazine at 0.70 kg ha-1, and bentazon at 1.4 kg ha-1 mixed with mesotrione at 0.11 kg ha-1 or topramezone at 0.02 kg ha-1. POST herbicide treatments were applied when corn had reached the V4 to V6 growth stages and the dominant weed species were no taller than 10 cm. The substitution of pyridate and bentazon for atrazine in tank mixes with mesotrione and topramezone did not significantly affect weed control for most species. Across all trial sites, the dominant and commonly occurring weeds (which included common lambsquarters (Chenopodium album), pigweeds (Amaranthus spp.), and common purslane (Portulaca oleracea)) were largely managed (>90% control at harvest) by the POST treatments. Exceptions included ivyleaf morningglory (Ipomoea hederacea) and fall panicum (Panicum dichotomiflorum) in NJ and Setaria spp. in NY. With respect to ivyleaf morningglory, tank mixes of pyridate (both rates) and bentazon with topramezone were less effective (68 to 80% control) than with mesotrione (100% control). Morningglory control was 100% when atrazine was included in the tank, regardless of HPPD active ingredient. Fall panicum was only 85 to 94% controlled in all mesotrione containing programs compared to 100% for topramezone programs. POST herbicide injury (up to 28%) was only observed at the PA site with pyridate at 0.71 kg ha-1 being more injurious to 'Serendipity' sweet corn than at the 0.35 kg ha-1 rate, regardless of HPPD tank mix partners. In NY and NJ, chlorosis and stunting were noted in the nontreated plots as the impacts of weed competition on sweet corn became visibly apparent. Yields were only affected at the NJ site, where POST applications reduced weed pressure and subsequently increased yields. The NY site trended, numerically, towards increased yields following POST treatments, but raccoon damage forced an early harvest of the trial. Yields at the PA site may have been impacted by dry weather conditions during ear fill. Pyridate, under the conditions of these trials, was a safe and effective tank-mix partner for mesotrione and topramezone in sweet corn.

PRE-TRANSPLANT APPLIED FLUMIOXAZIN, OXYFLUORFEN, AND SULFENTRAZONE EFFICACY AND SAFETY IN CABBAGE AND BROCCOLI IN NEW YORK AND NEW JERSEY.

Lynn M. Sosnoskie*[1], Thierry E. Besancon[2]; [1]Cornell University, Geneva, NY, [2]Rutgers University, Chatsworth, NJ (3)

Weed management remains one of the highest research priorities for cabbage producers. Weeds that evade control can compete, directly, with the cabbage crop for water and nutrients, resulting in yield loss. Weeds can also interfere with production operations, indirectly, by impeding harvest. Weed seeds can contaminate cabbage heads reducing their quality and increasing the labor required to meet grade standards. Weeds of the mustard family, such as Shepherd's purse, field pennycress and wormseed mustard, can serve as alternate hosts for the pathogens that cause Alternaria leaf spot, bacterial black rot, and club root diseases. The limited number of registered products and their narrow spectrums of control can result in significant in-season escapes and require the need for costly hand-weeding. Novel herbicide screens in NY have included the evaluation of flumioxazin over-the-top; in 2023 trials were established to evaluate the efficacy and safety of the product PRE-transplant. Research was conducted at the Cornell AgriTech Experiment Station in Geneva (NY) and the Rutgers Agricultural Research and Extension Center in Bridgeton (NJ) in late summer planted cabbage ("Padoc" and "Xtreme Vantage"). Trials were established on a Honeoye loam (fine, loamy, mixed, semiactive, mesic Glossic Hapludalfs, with 38% sand, 44% silt, 18% clay, 2.5% OM content, and a pH of 6.3) in NY and a Chillum silty loam (fine-silty, mixed, semiactive, mesic Typic Hapludults, with 54% sand, 28% silt, 18% clay, 2.4% OM content and a pH of 5.7) in NJ. Treatments included a non-treated check and PRE-transplant applications of oxyfluorfen at 0.2 and 0.56 kg ai ha^{-1}, sulfentrazone at 0.16 and 0.23 kg ai ha^{-1}, and flumioxain at 0.07 and 0.14 kg ai ha^{-1}. Transplants were set and herbicides were activated with at least 1.3 cm of irrigation water. Weed cover, weed control, and crop injury data were assessed for up to 5 weeks after transplanting (WAT) and at commercial harvest (NJ only). Weed and crop biomass assessments were also made. In NJ, all herbicide treatments effectively suppressed (< 1g total weed biomass per 0.25 m^2) the dominant weed species (e.g., CHEAL, ELEIN, AMARE, MOLVE) relative to the untreated check (34g total weed biomass per 0.25 m^2). In NY, all herbicides were largely effective at controlling (< 8g total weed biomass per 0.25 m^2) Polygonum spp. and THLAR, except for the sulfentrazone treatments (8 g and 19 g biomass per 0.25 m^2); the non-treated check averaged 41g total weed biomass per 0.25 m^2. Crop injury (e.g., chlorosis, necrosis, and stunting) was more pronounced in NJ as compared to NY, particularly in the flumioxazin treated plots. In NY, flumioxazin at 0.07 and 0.14 kg ai ha^{-1} stunted cabbage 15% and 25%, respectively, at 5 WAT. Only the high rate of flumioxazin impacted cabbage biomass (30% reduction in mean plant weight) in NY. In NJ, flumioxazin at 0.07 and 0.14 kg ai ha^{-1} stunted cabbage 84% and 96%, respectively, at 5 WAT; plots receiving the highest rate of flumioxazin also experienced significant stand loss. NJ cabbage yield was directly impacted by herbicide injury; flumioxazin treatments reduced head weights 80% to 90% relative to the non-treated check and the oxyfluorfen standard. Although the results from our trials indicate an impact of environment on injury potential, flumioxazin PRE-transplant does not appear to be an acceptable use pattern for cabbage.

EVALUATION OF THE AUTONOMOUS NAIO OZ ROBOT FOR WEED CONTROL IN SEEDED CORN AND TRANSPLANTED CABBAGE CROPS.

Wesley M. Bouchelle*[1], Carrie Mansue[1], Lynn M. Sosnoskie[2], Thierry E. Besancon[1]; [1]Rutgers University, Chatsworth, NJ, [2]Cornell University, Geneva, NY (4)

Although herbicides are a significant component of annual specialty crop production systems, the increased number of herbicide resistant weeds, the limited number of registered products, labor shortages, changing public perceptions about pesticide use, and a regulatory environment that is becoming more restrictive necessitates the evaluation of novel weed control technology. Autonomous weeders, such as the Naio Oz, have the potential to become important weed control tools for many agricultural systems. Autonomous robot weeders can reduce the time spent hand weeding, hoeing, and reduce or eliminate herbicide inputs. This technology could be used for organic growers, controlled environment agriculture, and in specialty crops where herbicide registrations are restricted. Trials were conducted in 2023 at Rutgers Agricultural Research and Extension Center in Bridgeton, New Jersey to test the effectiveness of the Naio Oz autonomous cultivator in direct seeded field corn and transplanted cabbage crops. When comparing the Naio Oz to standard herbicide treatments in both the corn and cabbage crops the weed biomass reduction compared to the control showed a significant decrease. When cultivation was used in corn it took two cultivation passes to achieve a 95% reduction in weed biomass in comparison to the standard post emergence herbicide treatment with a 99% reduction in weed biomass. When using Naio Oz cultivator for weed control in transplanted cabbage it took at least four passes of cultivation when no herbicides were used to achieve similar weed control as the standard herbicide program. When bensulide was applied pre-emergence with cultivation as post emergence control the number of passes of the Naio Oz was reduced by half to achieve the same reduction in weed biomass as the standard herbicide treatment and respectfully showed an 88% and 98% reduction in weed biomass. The use of Naio Oz robot cultivator shows potential to become an effective weed control tool for specialty crops.

WEED CONTROL AND FIELD CORN TOLERANCE WITH SOIL ADJUVANTS MIXED WITH RESIDUAL HERBICIDE AT PLANTING.

Lynn M. Sosnoskie[1], Wesley M. Bouchelle[2], Thierry E. Besancon*[2]; [1]Cornell University, Geneva, NY, [2]Rutgers University, Chatsworth, NJ (5)

Soil adjuvants are advertised as tools for improving the distribution and penetration of residual herbicides, which can, in turn, enhance the suppression of emerging weeds. It has also been suggested that, under dry situations, the use of soil adjuvants can improve product persistence until effective activation conditions can be achieved. Under conditions of heavy rainfall, soil adjuvants may be able to stabilize the soil to prevent herbicide loss by erosion or leaching, which can result in reduced weed suppression. To date, the amount of peer-reviewed, published literature around the performance of soil adjuvants is minimal. Field trials were conducted on field corn in spring 2023 at the Cornell AgriTech Experiment Station in Geneva, NY, and at the Rutgers Agricultural Research and Extension Center in Bridgeton, NJ, to evaluate three soil adjuvants (ORO-RZ at 2.3 L ha^{-1}, Grounded at 4.6 L ha^{-1}, and Infuse at 4.6 L ha^{-1}) tank mixed with either atrazine at 454 g ai ha^{-1}, S-metolachlor at 1.07 kg ai ha^{-1}, or pendimethalin at 1.06 kg ai ha^{-1} applied immediately after seeding. Treatments were arranged as a two factors (herbicide and soil adjuvant) factorial arrangement in a randomized complete block design with four replications. Averaged over soil adjuvants, weed coverage was greater with atrazine (3% and 13% at 21 and 48 DAT, respectively) than with *S*-metolachlor or pendimethalin (<1% and 9% averaged over herbicide at 21 and 48 DAT, respectively). The addition of soil adjuvants to herbicide spray mixtures had no effect on visual evaluations of weed coverage at 21 or 48 DAT. At both locations, greater densities of dicotyledonous weeds were noted following *S*-metolachlor (44 to 72 plants m^{-2}) than atrazine (8 to 20 plants m^{-2}). Conversely, lower grass densities were observed with pendimethalin and *S*-metolachlor (5 to 55 plants m^{-2}) than atrazine (30 to 1,190 plants m^{-2}). Overall, and relative to the untreated control, ORO-RZ was more effective at reducing weed density (79%) than Grounded (61%) or Infuse (67%). Total weed biomass was significantly higher following atrazine (480 g m^{-2}) than *S*-metolachlor or pendimethalin application (<200 g m^{-2}). Soil adjuvants impacted total weed biomass; across all herbicides, lower weed biomass was observed when ORO-R was included in the spray tank (63 g m^{-2}) compared to Infuse or the lack of soil adjuvant mixing partner (210 and 175 g m^{-2}, respectively). Corn stalk biomass 48 DAT was not affected by herbicides nor soil adjuvants. Because all active ingredients in this study are not water soluble, they are already formulated with surfactant to solubilize them in solution. Thus, the addition of soil adjuvants will not likely improve their deposition and dispersion.

ELUCIDATING WEED COMPETITION AND MANAGEMENT IN SOUTH CAROLINA HEMP.

Lynn M. Sosnoskie[1], Matthew A. Cutulle[2], Harrison T. Campbell*[2]; [1]Cornell University, Geneva, NY, [2]Clemson University, Charleston, SC (6)

Field experiments were conducted in 2022 and 2023 investigating the impact of weed-free intervals in floral hemp (*Cannabis sativa L.*) production in coastal South Carolina; furthermore, various weed control methods were examined in a separate trial also taking place in 2022 and 2023 with the aim of determining potential impacts on floral production as well as best weed control practices for various grower methods. The hemp cultivars examined in these studies included the following: Cherry Wine, Bubbatonic, Joey, and Rincon. Dry flower yields, among other parameters, of 3 cultivars per trial year (Cherry Wine, Bubbatonic, and Joey in 2022 and with Rincon replacing Joey in 2023) were assessed in relation to various weed-free intervals (0, 1, 2, 4, and 6 weeks weed-free and season long weed-free). The experiment was constructed as a strip-plot design with 4 replications. Weeding was accomplished by manually hoeing around each plant so that the entire area of the raised bed plot was weed free. When including all cultivars and pooling across trial years a significant increase in mean floral yield existed with 2, 4, 6, and season long weed-free intervals when compared to 0 weeks weed-free. Furthermore, a significant decrease in dry weed biomass existed when comparing the same weed-free intervals to 0 weeks weed-free. In a separate trial, weed management was assessed in relation to the following treatments: black plastic mulch, flaming at 84 L ha^{-1} and 140 L ha^{-1}, cultivation, and untreated bare ground check. The various treatments and the two cultivars (Cherry Wine and Bubbatonic) were arranged in a randomized split-plot design with four replications. Floral yield, among other parameters, was examined in relation to these treatments. Pooling weed control data across both trial years, carpetweed (*Mollugo verticillate*) and Palmer amaranth (*Amaranthus palmeri*) were significantly controlled by cultivation and both flaming rates. Yellow nutsedge (*Cyperus esculentus*) control was significantly increased by black plastic mulch and cultivation, respectively, when compared to all other treatments, while barnyardgrass (*Echinochloa crus-galli*) control was significantly increased with the addition of black plastic mulch when compared to all other treatments. Based on each of these studies, growers in coastal South Carolina will benefit from aiming to extend weed control four to six weeks after transplanting hemp, when possible, and could potentially see yields comparable to maintaining a weed-free setting throughout the entire growing season. The utilization of a plastic mulch barrier is also encouraged based on this study and will help to control problem weeds, such as barnyardgrass and yellow nutsedge, while flaming and cultivation will help to control less troublesome weeds when plastic mulch is not a viable option.

IR-4 PROJECT: SUCCESS AND BENEFITS TO SPECIALTY CROP GROWERS.

Roger B. Batts*, Alice Axtell, Jaimin Patel, Jerry Baron, Debbie Carpenter, Hannah Ross; IR-4 Project HQ, NC State University, Raleigh, NC (7)

IR-4 Project: Success and Benefits to Specialty Crop Growers. Roger B. Batts, Jaimin Patel, Alice Axtell, Jerry Baron, Debbie Carpenter, and Hannah Ross. IR-4 Project, NC State University, Raleigh, NC Last year, 2023, marked the 60-year milestone of the IR-4 Project helping growers/farmers of specialty crops (fruits, vegetables, nuts, herbs, spices, ornamentals and other horticultural crops) gain access to registrations of safe and effective chemical and bio-based herbicides, fungicides and insecticides. The IR-4 Project remains relevant because the crop protection industry focuses their research and development efforts on products that provide large sales that yield adequate return on investment. They shy away from specialty crops because of the cost of development of the data required for registration. The IR-4 Project fills this gap of developing data utilizing a network of public sector researchers (University and USDA-ARS) with expertise in pest management and analytical chemistry. Since its inception, IR-4 has secured over 23,000 registrations of crop protection products in food crops and over 56,000 uses in ornamental crops under its Environmental Horticulture program. IR-4 contributes nearly $9 billion to the annual US gross domestic product, according to a 2022 report by the Michigan State University Center for Economic Analysis. IR-4 activities include, but are not limited to: Food Crop Program: Facilitates regulatory approval of pest management solutions for specialty food crops through three research platforms: · - Residue Studies - determining the amount of chemical pesticide remaining in the crop at harvest; - Product Performance - developing data to show that a potential use of a pesticide is safe and effective; and - Integrated Solutions - utilizing multiple tactics including chemical pesticides, biopesticides, emerging technologies and other tools in combination to manage critical pests. Environmental Horticulture Program: Supports regulatory approval of pest management solutions for environmental horticultural crops including landscape/nursery plants, cut flowers and more. Biopesticide Regulatory Support: Aids in development and registration of bio-based pesticides for use on specialty crops. International Activities: Facilitates the international harmonization of Maximum Residue Levels (MRLs), supporting U.S. specialty crop growers in accessing export markets. IR-4 also helps build capacity of global minor use programs and collaborates with international partner organizations. For more information, please visit our website: https://www.ir4project.org/

WEED AND CHRISTMAS TREE RESPONSE TO TOPRAMEZONE HERBICIDE.

Jatinder S. Aulakh*[1], Vipan Kumar[2]; [1]The Connecticut Agricultural Experiment Station, Windsor, CT, [2]Cornell University, Ithaca, NY (8)

Field and container experiments were conducted to determine the effectiveness of topramezone, applied POST, for weed control and the safety of Christmas trees. For the container experiment, the bare-rooted seedlings (2+1) of balsam fir [*Abies balsamea* (L.) Mill. var. *balsamea*], Canaan fir [*Abies balsamea* (L.) Mill. var. *phanerolepis* Fernald], Colorado blue spruce (*Picea pungens* Engelm.), Norway spruce [*Picea abies* (L.) Karst.], and white pine (*Pinus strobus* L.) were planted into 2.9 L containers in the spring of 2019 at the Valley laboratory of the Connecticut Agricultural Experiment Station in Windsor, CT. The planting media consisted of composted woodchips, field soil, and commercial potting mixture (3:1:1 by volume). For the field experiment, the bare-rooted Fraser fir [*Abies fraseri* (Pursh) Poir.] seedlings (3+1) were transplanted in the spring of 2017 at a commercial Christmas tree farm in Enfield, CT. Topramezone herbicide was applied at 0, 98, 196, or 392 g ai ha^{-1} over the top (OTT) of tested Christmas tree species in both experiments in the spring of 2020 and 2021. A methylated seed oil surfactant was added (1% v/v) to all topramezone treatments. Treatments were applied 2 or 4 wk after bud-break in the container experiment and approximately 5 wk after bud-break in the field experiment. A compressed CO_2 backpack sprayer equipped with a single TeeJet 8002 nozzle delivering 187 L ha^{-1} at 207 kPa and 3.5 kph was used. Christmas tree injury (chlorosis, necrosis, and stunting) and weed control (field experiment) were evaluated at 2, 4, and 8 wk after treatment (WAT) using a 0-100% rating scale. Results from both the container and field experiments indicated that topramezone was highly safe for the tested Christmas tree species. The highest rate of 392 g ai ha^{-1} caused minor chlorotic injury on the field planted Fraser fir, only in 2020. The chlorotic injury was rated 8.2%, 3.3%, and 1.7% by 2, 4, and 8 WAT, respectively. In the field experiment, Canada horseweed [*Conyza canadensis* (L.) Cronq], Carolina horsenettle (*Solanum carolinense* L.), large crabgrass [*Digitaria sanguinalis* (L.) Scop.], and yellow foxtail [*Setaria pumila* (Poir.) Roem. & Schult.] were controlled 60% to 90% by 4 WAT. By 8 WAT, both Canada horseweed and large crabgrass were controlled >85%, regardless of the topramezone rate. Carolina horse nettle control varied from 35% to 60% whereas the yellow foxtail showed almost complete recovery (<15% control). These results suggest that topramezone is a highly safe POST herbicide on tested Christmas tree species and can provide moderate control or suppression of some of the annual and perennial weed species tested in this study.

CONTROL OF GLYPHOSATE-RESISTANT COMMON WATERHEMP (AMARANTHUS TUBERCULATUS) IN 2,4-D/GLYPHOSATE/GLUFOSINATE-RESISTANT SOYBEANS IN NEW YORK.

Vipan Kumar*[1], Mike Stanyard[2], Mike Hunter[3], Lynn M. Sosnoskie[4], Jatinder S. Aulakh[5]; [1]Cornell University, Ithaca, NY, [2]Cornell University, Newark, NY, [3]Cornell University, Watertown, NY, [4]Cornell University, Geneva, NY, [5]The Connecticut Agricultural Experiment Station, Windsor, CT (9)

Glyphosate-resistant (GR) common waterhemp has recently been identified in the northeastern United States, including New York and Connecticut. The evolution of GR waterhemp is a serious concern for producers in the region and warrants alternative herbicide strategies for its management. The main objective of this research was to determine the effectiveness of various preemergence (PRE) and postemergence (POST) herbicides for GR common waterhemp control in 2,4-D/glyphosate/glufosinate-resistant (Enlist E3) soybeans. To fulfill this objective, two separate studies were established during the summer of 2023 on a grower field in Seneca County, NY. The experimental field site was under corn-soybean rotation for >5 yrs with a known history of GR common waterhemp seedbank. An Enlist E3 soybean variety was planted on May 21, 2023. Herbicide programs, including PRE alone and PRE followed by (*fb*) early POST were tested in study 1, whereas programs including early POST alone or early POST *fb* mid POST were tested in study 2. Both studies were conducted in a randomized complete block design with three replications. Results from study # 1 indicated that PRE-alone applications of chlorimuron + flumioxazin + metribuzin, imazethapyr + flumioxazin + metribuzin, acetochlor + fomesafen, and s-metolachlor + metribuzin at field-use rates provided 63 to 87% control of GR common waterhemp 25 days after PRE (DAPRE); however, control did not exceed 32% with any of those treatments at 42 days after mid POST (DAMPOST). In contrast, all PRE programs *fb* an early POST treatment of glufosinate + 2,4-D consistently provided excellent season-long control (82 to 94%) and significantly reduced shoot biomass (up to 96%) of GR common waterhemp. In study # 2, the GR common waterhemp was controlled =93% at 28 DAMPOST with either a single early POST application of 2,4-D + (acetochlor + fomesafen), 2,4-D + glufosinate + (acetochlor + fomesafen) or sequential POST applications of 2,4-D + (acetochlor + fomesafen) *fb* 2,4-D, glufosinate + (acetochlor + fomesafen) *fb* glufosinate, 2,4-D + (acetochlor + fomesafen) *fb* glufosinate, glufosinate + (acetochlor + fomesafen) *fb* 2,4-D or 2,4-D + glufosinate + (acetochlor + fomesafen) *fb* 2,4-D + glufosinate. All sequential POST programs significantly reduced (97 to 100%) biomass of GR waterhemp in study 2. Altogether, results suggest that effective alternative PRE *fb* POST or sequential POST herbicide programs tested in this research can help controlling GR common waterhemp in 2,4-D/glyphosate/glufosinate-resistant soybeans.

DOES PERENNIAL AND COVER CROP DIVERSITY INFLUENCE MICROBIALLY-MEDIATED WEED SEED MORTALITY?

Kara E. Eckert*, Carolyn Lowry, Eric Larson, Sharifa Crandall, Estelle Couradeau; Penn State University, University Park, PA (10)

Understanding the microbial communities that regulate weed seed mortality in the soil seedbank is a promising approach for early ecological control of annual, problematic weeds. Certain soil microbes are known to infect seeds in the soil, resulting in mortality, and act as important regulators of weedy populations. However, one major knowledge gap is whether certain crop communities can affect these populations of soil microbes and regulate weed seed populations in the soil seedbank. Currently, we have examined a perennial forage system to determine crop influence on weed seed mortality with microbial analyses soon to come. The perennial forage treatments (7) include monocultures and mixes of *Medicago sativa* (legume), *Dactylis glomerata* (grass), and *Cichorium intybus* (forb). Weed seeds (*Amaranthus powellii, Abutilon theophrasti*) are buried in mesh bags under crops for a minimum of 1.5 years, and then undergo viability and microbial testing. Neither perennial treatment nor species diversity has significant impact on weed seed mortality after burial for 1.5 years. However, there was a significant negative correlation between proportion of alfalfa in the mixture and *A. powellii* viability. Upcoming research will investigate (1) weed seed viability up to 2.5 years buried, (2) crop influence on the soil microbiome, (3) crop influence on the weed seed microbiome, and (4) correlation between weed seed microbiome and seed mortality.

ARE ORGANIC HERBICIDES EFFECTIVE FOR BURNDOWN PRIOR TO CROP ESTABLISHMENT?

Dwight Lingenfelter*; Penn State University, University Park, PA (11)

Weed management in organic cropping systems is extremely challenging. In most cases, it is accomplished with various tactics including tillage, cover crops, mulches, among others. Planting vegetable crops into an organic no-till system provides its own set of unique challenges and significantly reduces the number of weed control options. However, some growers are always looking for ways to make this system work. One option is to apply OMRI-approved herbicide products such as HomePlate, Axxe, or Scythe during the burndown phase prior to crop planting. However, reliable information is limited on the overall weed control effectiveness of these types of products. Initial university data indicates that high product use rates and high spray volumes will likely be necessary for adequate burndown of existing weeds. Thus, a field study was conducted to examine various OMRI approved herbicides to determine their effectiveness on burndown weed control as well as compare them to typical burndown herbicides. An experiment was conducted at the Russell E. Larson Agricultural Research Farm in Centre County in 2022. Several herbicide treatments including HomePlate (3, 6, and 9%, caprylic acid + capric acid), Axxe (8 and 13%, ammonium nonanoate), Scythe (5 and 10%, pelargonic acid), Roundup PowerMax (0.75 lb ae, glyphosate), and Liberty (0.59 lb ai, glufosinate) were evaluated in a randomized complete block design with three replications. Each plot was 5 x 30 feet and established in a fallow area with horseweed (*Conyza canadensis*) and other winter annual or perennial weeds. Treatments were applied on May 5. Standard flat fan spray nozzles (TeeJet XR 11003) and two spray volumes (35 & 70 gallons/acre) were used to maximize coverage on the weed foliage. Nu-Film P (OMRI-approved) adjuvant was included in some treatments. Weed control for all species present was evaluated on May 9 and May 20. Preliminary results from this study showed that none of the "organic herbicide" treatments in this study provided effective burndown weed control. In general, organic herbicides cause substantial leaf burning at first, but most weeds overcame this initial injury and thrived afterwards. Rating taken two weeks after application showed that none of the organic treatments provided greater than 70% control of the weeds present, with a few exceptions. Most ranged from 20-60% control. Highest labeled use rates were necessary to get better weed suppression. High spray volumes (70 gallons/acre) provided better coverage of weed foliage and tended to cause more leaf injury. Certain OMRI-approved adjuvants (e.g., Nu-Film P) slightly improved weed control in this study. And finally, weed control with Liberty and Roundup continued to increase over the evaluation period on most weed species.

EFFICACY OF TIAFENACIL +/- GLUFOSINATE ON SUMMER ANNUAL WEEDS.

Christopher D. Harlow*, Joe C. Neal; North Carolina State University, Raleigh, NC (12)

Weed control options that do not include glyphosate continue to be desired in landscape plantings. Tiafenacil and glufosinate are considered to be non-selective, but both have lower efficacy on some species compared to glyphosate. The present study was initiated to evaluate efficacies of tiafenacil and glufosinate, applied individually or combined in a tank-mix, on summer annual weeds in a landscape bed setting. The experiment was conducted in outdoor field research beds in 2022. Beds were cultivated on July 17, 2022, and herbicide treatments were applied on August 2, 2022. At the time of application, multiple pigweed species (*Amaranthus* spp.), goosegrass (*Eleusine indica)*, carpetweed (*Mollugo verticillata L.*), spotted spurge (*Euphorbia maculata*), and crabgrass species (*Digitaria* spp.) were well-established in the beds. Treatments included tiafenacil at 0.375, 0.75 or 1.5 lb ai/A without adjuvant, tiafenacil at 1.5 lb ai/A + 1% MSO; glufosinate at 0.5, 1.0 or 1.5 lb ai/A; and factorial combinations of tiafenacil at 0.375, 0.75 or 1.5 lb ai/A with glufosinate at 0.5 or 1 lb ai/A. Nontreated check plots were included for comparison. Percent weed control was visually evaluated 3, 8, 17, and 29 days after application using a 0 to 10 scale where 0 = no control (indistinguishable from nontreated plots) and 10 = 100% mortality. Additionally, percent living (green) ground cover in plots was estimated. The addition of an adjuvant (MSO) dramatically improved tiafenacil efficacy. Both carpetweed and spurge were controlled over 80% by tiafenacil + MSO but were not controlled by tiafenacil alone. Pigweed control ranged from 43 to 63% with tiafenacil alone but was over 90% with the addition of MSO. Summer annual grass control was 65% with tiafenacil at 1.5 lb ai/A + 1% MSO, while no grass control was observed in the absence of MSO. Glufosinate at 0.5 lb ai/A controlled most dicot weeds but was less effective on grasses. Increasing the dose of glufosinate to 1.0 lb ai/A or 1.5 lb ai/A provided much better control of grasses and nearly complete control of dicot weeds. The addition of tiafenacil to glufosinate did not significantly improve weed control over glufosinate applied alone. However, no adjuvant was included in these tank-mix combinations. Given the improved efficacy of tiafenacil at 1.5 lb ai/A with the addition of MSO, it would be useful to repeat the tank-mix evaluations with the addition of MSO to all treatments.

OPTIONS FOR HORSENETTLE CONTROL IN LANDSCAPE AND NURSERY SITES.

Joseph C. Neal*, Christopher D. Harlow; North Carolina State University, Raleigh, NC (13)

Perennial broadleaf weeds are challenging to manage in field nurseries and landscape plantings. Directed applications of glyphosate are generally effective on many weeds but unsatisfactory control of horsenettle (*Solanum carolinense* L.) is often observed. Several postemergence herbicides were compared for control of horsenettle. The experiment was conducted in outdoor containers in 2022 and repeated in 2023. Horsenettle roots were collected from local infestations in March of each year. Root pieces were potted to 4-L containers using a pine bark substrate and top-dressed with slow-release fertilizer. Plants were grown for two months and pruned periodically to about 6 inches in height to encourage branching and root sprouting. Herbicides were applied on July 6, 2022, and on June 28, 2023. Contact-action herbicides were re-applied at 2 weeks; systemic herbicides were reapplied at 4 weeks. Herbicides evaluated were 2,4-D amine, bentazon, clopyralid, flumioxazin + pyroxasulfone, glufosinate, glyphosate, rimsulfuron, tiafenacil, topramezone, triclopyr amine, and combinations of glufosinate with 2,4-D or topramezone. Each was applied at the manufacturers' recommended doses and with adjuvants as specified on the product labels. Herbicides were applied with a CO_2-pressurized sprayer equipped with TeeJet 8008 flat fan nozzles and calibrated to deliver 50 GPA. Percent horsenettle control was visually evaluated for 8 weeks. Results from 2022 and 2023 were similar. Glyphosate at 2 lb ai/A provided 38% and 76% control in 2022 and 2023, respectively. Triclopyr, flumioxazin + pyroxasulfone, glufosinate, 2,4-D amine, and topramezone controlled horsenettle 95, 87, 80, 79, and 79%, respectively, each significantly greater than glyphosate. Clopyralid caused typical synthetic auxin injury symptoms and some stunting, but plants survived and continued growth. Tiafenacil caused some initial leaf necrosis, but plants resumed growth soon after treatment. Rimsulfuron caused mild and temporary chlorosis but no reduction in growth. Bentazon was ineffective. Horsenettle control with a single application of 2,4-D or topramezone tank mixed with glufosinate did not control the weed as well as two applications of the herbicides applied individually. There was a suggestion of potential antagonism from 2,4-D + glufosinate applications that should be investigated. Results from this test suggest that triclopyr, flumioxazin + pyroxasulfone, glufosinate, 2,4-D amine, or topramezone have the potential to control horsenettle better than glyphosate, but results need to be confirmed by tests in established natural stands of horsenettle.

INCENTIVIZING PRODUCTION OF CLIMATE SMART LEAFY GREENS IN SOUTH CAROLINA.

Matthew A. Cutulle*[1], Brian Ward[1], Kari Buck[2], Lacy Barnette[3], Jhessye Moore-Thomas[3], Kelly Flynn[2], Paula Agudelo[2]; [1]Clemson University, Charleston, SC, [2]Clemson University, Clemson, SC, [3]Clemson, Columbia, SC (14)

South Carolina agriculture face unique challenges with soils that are highly degraded. Most of these soils have organic matter content less than 1% and low water holding capacity that impair agroecosystem sustainability and climate resilience. Most Southeastern soils where leafy greens are grown contain a subsurface compacted zone (hardpan) limiting root penetration, which predisposes crops to drought stress, reduce yields, and impair climate resilience of cropping systems. Intense and poorly distributed rainfall and high temperatures contribute to soil degradation, and intensive tillage further exacerbate the issue. To manage soil compaction and control weeds farmers practice heavy tillage, which adds to production costs, depletes organic matter, and leaves the soil prone to re-compaction. To address these challenges, we have incentivized conservation management practices to decrease soil organic carbon losses while maintaining soil health, agricultural productivity and climate resilience in leafy green productions systems. Practices that are promoted in by this project include cover crops, reduced tillage and mulching inclusive of organic material, biodegradable polymers and UV reactive mulch. Currently we have approximately 100 leafy greens growers enrolled in our climate smart practices incentive programs. In addition to soil health and greenhouse gas emission data collected from this study researchers will also profile the impact of climate smart practices on weed species composition over time.

DELAYED CEREAL RYE TERMINATION INFLUENCES WEED RECRUITMENT: A REGIONAL PERSPECTIVE.

Grant M. Hoffer*[1], John M. Wallace[2], Karla L. Gage[3], Mark J. VanGessel[4], John Lindquist[5], Wesley Everman[6], Erin Haramoto[7], Eugene P. Law[8], Nicholas T. Basinger[9], Eric J. Miller[10]; [1]Pennsylvania State University, University Park, PA, [2]Penn State University, University Park, PA, [3]Southern Illinois University Carbondale, Carbondale, IL, [4]University of Delaware, Georgetown, DE, [5]University of Nebraska, Lincoln, NE, [6]North Carolina State University, Raleigh, NC, [7]University of Kentucky, Lexington, KY, [8]University of Delaware, Beltsville, MD, [9]The University of Georgia, Athens, GA, [10]Southern Illinois University, Carbondale, IL (15)

Planting green is gaining interest among no-till soybean (*Glycine max*) producers as a means of increasing soil health and managing herbicide-resistant weeds but increases the complexity of crop management decisions. This experiment's objective is to inform regional- and site-specific Decision Support Tools (DSTs) for planting green by evaluating the effects of cereal rye (*Secale cereale L.*) termination timing on (1) increases in cereal rye biomass production, (2) broadleaf weed recruitment patterns at a POST herbicide application timing, (3) the efficacy of soil-applied residual herbicides, and (4) soybean yield. We expected that delayed termination of cereal rye would have a multiplicative effect on weed recruitment when combined with soil applied herbicides, but the magnitude of this effect would vary based on marginal gains in cereal rye biomass and weed species that occur across production regions. The experimental design was a 3x2 factorial arranged as a split-plot with four replicates at each site and two sites at each of the nine participating locations spanning multiple production regions (PA, MD, DE, NC, GA, KY, IL, NE). Main plot treatments were a fallow control, cereal rye terminated 14-21 days before soybean planting (14 DPP), and cereal rye terminated 1-3 days after soybean planting. Split plot treatments were the presence or absence of a PRE application of Fierce XLT (chlorimuron, flumioxazin, and pyroxasulfone mixture). Weed density data was collected 28 days after planting (28 DAP), the R1 soybean growth stage, and at post leaf-drop. Cover crop biomass was collected immediately preceding termination and soybean yields were measure in the middle two rows using a small plot combine. Based on linear mixed model regression, average cereal rye biomass 14-21 DPP was 2.9 Mg ha^{-1} and increased 70 kg ha^{-1} day^{-1} as termination was delayed. When compared to the weedy check (fallow control, no PRE) at the 28 DAP weed demographic census, the combination of terminated cereal rye 14 DPP and use of a PRE application significantly decreased weed recruitment in 13 of 16 sites. In the absence of the PRE application, planting green significantly decreased weed recruitment in 8 of 16 sites. The combination of planting green and PRE tactics resulted in significant weed suppression at 14 of 16 sites. Significant soybean yield differences between planting green occurred at only two locations (NE, IL), where planting green resulted in either reduced (NE) or increased (IL) yields relative to the fallow treatment. This experiment will be replicated for multiple years to fully characterize the impacts of planting green on weed and soybean management across a range of soil-environment conditions.

SMALL CARPETGRASS (ARTHRAXON HISPIDUS) IS SUSCEPTIBLE TO COMMONLY USED PRE AND POST NON-CROP HERBICIDES. Arthur E. Gover*; Penn State, University Park, PA (16)

ABSTRACT NOT AVAILABLE

MANAGING AN INVASIVE KNAPWEED (CENTAUREA NIGRANS) IN COOL-SEASON PASTURES.

Rakesh S. Chandran*; West Virginia University, Morgantown, WV (17)

Black knapweed (*Centaurea nigrans*) is an emerging invasive weed in Appalachian cool-season pastures. Field experiments were conducted in a pasture near the Sinks-of-Gandy, West Virginia, during 2022 and 2023 to determine the effect of fall-application of 7 broadleaf herbicides for their effectiveness to control this weed. Herbicide treatments and application rates (lb ai/A) consisted of aminopyralid+florpyrauxifen (0.083+0.035), 2,4-D + dicamba (0.5+1.4), aminopyralid+2,4-D (0.078+0.625), clopyralid (0.375), fluroxypyr+triclopyr (0.373+0.125), aminopyralid (0.078), and saflufenacil (0.044). All herbicide treatments except saflufenacil were applied with methylated seed oil (MSO) @ 1% vol/vol as a surfactanct; saflufenacil treatment contained a non-ionic surfactant @ 0.25% vol/vol. Treatments were applied on September 14, 2022 and weed control evaluations along with forage biomass were recorded nine months later (June 2023). All treatments except fluroxypyr+triclopyr and saflufenacil provided effective (>90%) black knapweed control. Forage biomass from the plots that received the effective treatments increased by 2 to 3 times greater than that from the untreated plots.

AN INTEGRATED APPROACH FOR THE RESTORATION OF TEMPERATE GRASSLANDS INVADED BY AN NASSELLA TRICHOTOMA. Talia J. Humphries*; Texas A&M University, College Station, TX (18)

ABSTRACT NOT AVAILABLE

2023 SURVEY RESULTS FOR THE MOST COMMON AND TROUBLESOME WEEDS IN GRASS CROPS, PASTURE AND TURF.

Cynthia Sias[1], Annu Kumari[2], Lee Van Wychen*[3]; [1]Virginia Tech, Blacksburg, VA, [2]Crop, Soil & Environmental Sciences Department, Auburn University, Auburn, AL, [3]Weed Science Society of America, Alexandria, VA (19)

The 2023 Weed Survey for the U.S. and Canada surveyed the most common and troublesome weeds in the following grass crops: 1) corn (*Zea mays*); 2) sorghum (*Sorghum bicolor*); 3) spring cereal grains; 4) winter cereal grains; 5) pastures, rangeland, or other hay; and 6) turf. Common weeds refer to the weeds you most frequently see while troublesome weeds are the most difficult to control but might not be widespread. There were 253 survey responses from the U.S. and Canada. In corn, the top five most common weeds were 1) common lambsquarters (*Chenopodium album*); 2) waterhemp (*Amaranthus tuberculatus*); 3) morningglory species (Ipomoea spp.); 4) Palmer amaranth (*Amaranthus palmeri); 5)* giant foxtail *(Setaria faberi).* The most troublesome weeds in corn were 1) waterhemp; 2) morningglory spp.; 3) Palmer amaranth; 4) johnsongrass (*Sorghum halepense*); and 5) kochia (*Bassia scoparia*). In sorghum, the top three most common weeds were 1) Palmer amaranth 2) johnsongrass; and 3) a tie among kochia; morningglory spp.; and pigweed spp. The top three most troublesome weeds were: 1) johnsongrass; 2) Palmer amaranth; and 3) kochia. In spring cereal grains, the top three most common weeds were: 1) a tie between common lambsquarters and wild oat (*Avena fatua*); and 3) kochia. The top three most troublesome weeds in spring cereal grains were 1) wild oat; 2) kochia; and 3) green foxtail (*Setaria viridis).* In winter cereal grains, the top five most common weeds were 1) henbit (*Lamium amplexicaule*); 2) common chickweed (*Stellaria media*); 3) downy brome (*Bromus tectorum*); 4) Italian ryegrass (*Lolium perenne ssp. Multiflorum*); and 5) annual bluegrass (*Poa annua*). The most troublesome weeds were: 1) downy brome; 2) a tie between horseweed (*Conyza canadensis*) and Italian ryegrass; 4) annual bluegrass; and 5) kochia. In pastures, rangeland, and other hay, the top five most common weeds were 1) Canada thistle (*Cirsium arvense*); 2) horsenettle (*Solanum carolinense*); 3) dandelion (*Taraxacum officinale*); and 4) a tie between downy brome and musk thistle (*Carduus nutans*). The most troublesome weeds were 1) Canada thistle; 2) leafy spurge (*Euphorbia esula*); 3) horsenettle; 4) downy brome; and 5) johnsongrass. In turf, the top five most common weeds were 1) dandelion; 2) annual bluegrass; 3) white clover (*Trifolium repense*); 4) smooth crabgrass (*Digitaria ischaemum*); and 5) goosegrass (*Eleusine indica*). The most troublesome weeds were: 1) a tie between annual bluegrass and bermudagrass (*Cynodon dactylon*); 3) goosegrass; 4) yellow nutsedge (*Cyperus esculentus*); and 5) dallisgrass (*Paspalum dilatatum*). Overall, the top five most common weeds among all grass crops were 1) common lambsquarters; 2) kochia; 3) dandelion; 4) Canada thistle; and 5) Palmer amaranth. The most troublesome weeds were: 1) kochia; 2) Canada thistle; 3) johnsongrass; 4) Palmer amaranth; and 5) annual bluegrass.

ADVANCING WEED SCIENCE RESEARCH, EXTENSION, AND EDUCATION: NIFA GRANTS AND PANEL REVIEWER OPPORTUNITIES.

Annu Kumari[1], Cynthia Sias[2], James Kells[3], Vijay Nandula[4], Lee Van Wychen*[5]; [1]Crop, Soil & Environmental Sciences Department, Auburn University, Auburn, AL, [2]Virginia Tech, Blacksburg, VA, [3]Michigan State University, East Lansing, MI, [4]USDA, Stoneville, MS, [5]Weed Science Society of America, Alexandria, VA (20)

The United States Department of Agriculture-National Institute of Food and Agriculture (USDA-NIFA) administers competitive research, extension, and education programs in support of US agriculture. Several of these programs fund research and extension projects related to weed science along with other pest disciplines. Traditionally, a large number of successful weed science-focused project awards came from the Crop Protection and Pest Management (https://www.nifa.usda.gov/grants/funding-opportunities/crop-protection-pest-management) and the Agricultural and Food Research Initiative Foundational and Applied Science's (AFRI FAS, https://www.nifa.usda.gov/grants/programs/agriculture-food-research-initiative-afri/afri-foundational-applied-science-program) Pests and Beneficial Species of Agricultural Production Systems programs. However, there are several other programs that could be relevant to weed scientists. Some of these include, AFRI FAS – Critical Agricultural Research and Extension (CARE), IR-4 (https://www.nifa.usda.gov/grants/funding-opportunities/minor-crop-pest-management-program-interregional-research-project-4-ir), Methyl Bromide Transition (MBT, https://www.nifa.usda.gov/grants/funding-opportunities/methyl-bromide-transition-program), organic programs - Organic Agriculture Research and Extension Initiative (OREI, https://www.nifa.usda.gov/grants/funding-opportunities/organic-agriculture-research-extension-initiative) and Organic Transitions (ORG, https://www.nifa.usda.gov/grants/funding-opportunities/integrated-research-education-extension-competitive-grants-program-0), and Specialty Crop Research Initiative (SCRI, https://www.nifa.usda.gov/grants/funding-opportunities/specialty-crop-research-initiative). Additionally, several AFRI FAS program area priorities are available for projects involving interdisciplinary research and extension. Weed scientists are strongly encouraged to consider submitting proposals to these USDA-NIFA competitive grant programs. Weed scientists, especially early- and mid-career academic scientists, are also encouraged to volunteer to serve on a USDA-NIFA proposal review panel. Weed science representation on review panels is important and serving on a review panel is a great way to improve proposal-writing skills and build professional networks. Click on this link https://prs.nifa.usda.gov/prs/volunteerPrep.do to volunteer. Volunteering simply places your name on a list of candidate reviewers. There is no commitment until you are contacted by a USDA-NIFA Panel Manager/Program Director and agree to serve on a specific panel. (Copy and paste the respective URL links in a new tab or window of your web browser)

A MUTATION IN A TRANSCRIPTION FACTOR IMPARTS NON-TARGET SITE HERBICIDE RESISTANCE IN RICE.

Srishti Gupta*; Colorado State University, Fort Collins, CO (21)

Chemical mutagenesis using ethyl methanesulfonate was performed to develop quizalofop-p ethyl (QPE) resistant lines. The TSR1 line had a mutation in acetyl-CoA carboxylase, the target site of QPE. On the other hand, the NTSR1 line had a novel mutation imparting non-target site resistance to group 1 herbicide. Genetic mapping identified a single nucleotide polymorphism in an uncharacterized putative zinc finger transcription factor. Transcription factors (TF) play important role in many biological processes like cell growth, division, and response to abiotic factors, but no mutation in TFs has been associated with herbicide resistance to date. The objective of this research is to understand the molecular mechanism through which a mutation in a transcription factor imparts herbicide resistance. The degree of resistance of TSR1 and NTSR1 was assessed with dose-response curve experiments using QPE, relative to a wild-type rice line. This experiment was done in the greenhouse and QPE was applied at doses ranging from 2.4 to 1,000 g/ha with 0.1% NIS to 3 weeks old plants. Injury rating and biomass were measured 21 days after treatment. The rate of QPE metabolism was measured to determine the basis of resistance in wild-type rice, NTSR1 rice and TSR1 rice collected 1,2,4 and 8 days after treatment using LC-MSMS analysis. This was followed by identifying putative metabolites in plant extracts using LC-MSMS. Putative NTSR genes that may be regulated by the mutated transcription factor and contribute to herbicide resistance were investigated using RNA Seq analysis. In greenhouse assays, the NSTR1 and TSR1 lines were 1.9 and 2.4 more tolerant to QPE than the wildtype. On the other hand, a TSR1/NTSR1 double mutant line was 6.8 times more tolerant to QPE than wildtype.

CROP AND WEED DETECTION USING MACHINE LEARNING.

Rutvij S. Wamanse*[1], Dhiraj Srivastava[2], Fatemeh Esmaeilbeiki[3], Vijay Singh[2]; [1]Eastern Shore Agricultural Research and Extension Center, Virginia Tech, Painter, VA, [2]Virginia Tech, Painter, VA, [3]Virginia Tech, Blacksburg, VA (22)

Weeds can cause significant yield losses in cropping systems if not monitored and managed at early stage. One of the efficient ways of monitoring weeds is the use of Unmanned Aerial Systems (UAS) combined with advanced machine learning approaches. Recent studies have indicated that identifying weeds on a large-scale agricultural field require UAS-based image data under various weather conditions to capture the variation. The acquired image dataset are then processed using advanced machine learning algorithms. Robust machine learning models can be created by rigorously training them on high-quality image data, a resource which is currently lacking in weed science. The current research focuses on state-of-the-art object detection models to detect Common ragweed (Ambrosia artemisiifolia) in soybean (Glycine max). The image dataset was collected in 2021 at Painter, Virginia, where DJI M-300 drone was flown at a height of 12 m at various growth stages of common ragweed and soybean. In order to feed classification models, a collection of 500*500 pixels were collected from the orthomosaic image and the image data was annotated using the VGG Image annotator (VIA). Several model architectures using the most recent YOLOv8 series were used to detect weeds. Metrics like precision and recall were used to compare the models. To create realistic images of weeds, Progressive Generative Adversarial Networks (GANs) were utilized. With a prediction speed of 0.854 ms and a precision of 82.2%, YOLOv8 nano (YOLOv8n) was found to be the fastest model, while YOLOv8 extra-large (YOLOv8x) was the best model with a precision of 97.19%. According to the study's findings, the YOLOv8 model has potential for site-specific operations in real-time operations. Subsequent research efforts will focus on creating lightweight machine learning models with high accuracy for weed prediction, which can be deployed on edge devices and better generative AI models can be used to create hyper-realistic looking weed images to improve the training set.

EVALUATING REDUCED TILLAGE WITH A LIVING MULCH AS AN IPM TOOL IN BROCCOLI PRODUCTION.

Carrie Mansue*[1], Dwayne D. Joseph[2], Wesley M. Bouchelle[1], Cerruti Rr Hooks[3], Thierry E. Besancon[1]; [1]Rutgers University, Chatsworth, NJ, [2]University of Maryland, Chestertown, MD, [3]University of Maryland, College Park, MD (23)

The United States is the 3rd largest producer of broccoli in the world. With the increase production of broccoli over the past decades, this can lead to enhance pest status of some herbivores such as cabbage loopers as well as weed control problems. This project's objective was to examine the effects of planting living-mulch within the rows of broccoli plants to see if there was an effect on pests, weed density, weed biomass, and crop growth and yield. Results show that insect counts did decrease where broccoli plants were surrounded by the living-mulch. On average, broccoli plants planted in the bare-ground had 6 caterpillars per 24 plants per plot compared to 5 in the living-mulch throughout the trial. Data for the weed biomass at harvest results showed a decrease in weed biomass for plots containing the living much. Living-mulch on average had a dry weight of 2 g m^{-2} of weed biomass compared to 16 g m^{-2} in bare-ground. However, living-mulch plots showed a decrease in harvest yield of broccoli compared to bare-ground plots. On average the total yield of bare-ground plots was 24 kg m^{-2} per plot compared to 12 kg m^{-2} in living-mulch plots. We can conclude from our first year of data, that using living-mulch shows some potential towards controlling caterpillars and weeds. However, next year's trial spacing for living-mulch plots may need to be wider to see if this can help with harvest yield.

HERBICIDE RESISTANT REDROOT PIGWEED POPULATIONS IN NORTH CAROLINA.

Ronel J. Argueta*; North Carolina State University, Raleigh, NC (24)

Farms have limited chemical options to control *Amaranthus spp.* in conventional soybeans, relying mainly on acetolactate synthase (ALS)-inhibting and protoporphyrinogen oxidase (PPO)-inhibiting herbicides. In 2019 (Camden County) and 2020 (Pasquotank County), complaints of control failures with acetolactate synthase- and protoporphyrinogen oxidase-inhibiting herbicides on *Amaranthus retroflexus* L. (redroot pigweed) were reported by two different farmers in North Carolina. Greenhouse dose-response assays provided the Camden and Pasquotank County *Amaranthus retroflexus* populations survived lethal doses of ALS- and PPO-inhibiting herbicides. Greenhouse dose-response assays were conducted with imazethapyr, thifensulfuron-methyl, and fomesafen on both of the putative herbicide-resistant *Amaranthus retroflexus* populations. Two-three putative herbicide-susceptible *Amaranthus retroflexus* (Wake and Yadkin [A&B] County) were included in each dose-response assay as well. Lethal doses that control 50% of the population (LD_{50}) were compared across the putative herbicide-resistant and – susceptible *Amaranthus retroflexus* populations to derive resistance ratios (LD_{50} R/ LD_{50} S). The lactofen LD_{50} on the Camden County Amaranthus retroflexus population was 1709 g ai ha^{-1} (8.6x the maximum labeled rate) which resulted in resistance ratios of 143 (Wake County) and 61 (Yadkin [A] County). These results provide evidence of the evolution of ALS-, PPO-inhibiting and multiple herbicide-resistant *A. retroflexus* populations in North Carolina; representing the first confirmed case of PPO-inhibiting herbicide-resistant *A. retroflexus* in the United States. We confirm the evolution of ALS- and PPO-inhibiting herbicide resistance two North Carolina populations of *Amaranthus retroflexus.* Keywords: *ALS*; herbicide resistance; *PPO*; redroot pigweed; *Amaranthus retroflexus*; weed management.

INVESTIGATING EFFECTS OF MILKWEED ESTABLISHMENT STRATEGY ON RHIZOSPHERE COMMUNITY ASSEMBLY.

Linnea C. Smith*[1], Antonio DiTommaso[1], Jenny Kao-Kniffin[2]; [1]Soil and Crop Sciences Section, School of Integrative Plant Science, Cornell University, Ithaca, NY, [2]Horticulture Section, School of Integrative Plant Science, Cornell University, Ithaca, NY (25)

Milkweeds such as common milkweed (*Asclepias syriaca* L.) provide critical habitat for the monarch butterfly (*Danaus plexippus*), whose population has declined concurrently with that of milkweed in the last few decades. Milkweed is the sole food source for monarch caterpillars, making it a vital species for increasing populations of this charismatic pollinator species. Milkweed latex contains cardenolides, toxic steroidal compounds that are consumed by monarch caterpillars and constitute an important predation defense mechanism for monarchs. The importance of milkweeds is widely recognized, and institutions such as the USDA NRCS have programs providing financial incentives for farmers to establish milkweed stands on their land. Planting large stands of milkweeds and other native wildflower species is difficult, however, without expensive equipment unavailable to most farmers. This challenge led to the development of Multi-Seed Zea Pelleting (MSZP), a technique to package milkweed seeds into a seed pellet the size and shape of a corn seed, enabling planting with a standard corn drill. The present study will investigate the effect of the pelleting process on the establishment of the belowground microbial community around the milkweed. Specifically, we hypothesize that the addition of maltodextrin (a pellet component) to the soil will lead to a soil carbon priming effect favoring bacteria over fungi. A prior study has shown that the fungal species colonizing the milkweed affect the latex production of the plant, which could have consequences for the caterpillars that rely upon it. Therefore, understanding how MSZP affect the soil microbial community is relevant to the ultimate goal of milkweed establishment. We will perform a pot experiment from November 2023 – January 2024 assessing the soil microbial community structure, soil respiration, milkweed latex production, and cardenolide content in pots containing only the pellet mixture, free-planted seeds, or MSZPs. Future studies will incorporate soil legacy effects, investigating possible differential milkweed establishment on soils on which corn, wheat, or soy was previously grown. Understanding how MSZPs and soil legacy effects alter soil microbial communities, and the possibility of cascading effects on milkweed cardenolide content, will allow us to optimize MSZP technology to establish robust milkweed stands on participating farms.

MULTIPLE HERBICIDE-RESISTANT ITALIAN RYEGRASS IN VIRGINIA.

Milos Viric*[1], Akashdeep Singh Brar[1], Vipin Kumar[2], Michael L. Flessner[3], Vijay Singh[1]; [1]Virginia Tech, Painter, VA, [2]University of Nebraska-Lincoln, Lincoln, NE, [3]Virginia Tech, Blacksburg, VA (26)

Italian ryegrass is one of the most troublesome weeds in the wheat production system in Virginia. In the summer of 2020 and 2022, field surveys were conducted and extension agents from various counties were contacted to provide samples to document the distribution of herbicide-resistant *Lolium* spp. infesting winter wheat production fields in the region. A total of 32 samples were collected, dried, threshed and initially screened in a greenhouse for sensitivity to diclofop-methyl (516 g ai ha^{-1}), pinoxaden (59.43 g ai ha^{-1}), mesosulfuron (17.5 g ai ha^{-1}), pyroxsulam (17.94 g ai ha^{-1}), glyphosate (1032 g ai ha^{-1}), and pyroxasulfone (102.97 g ai ha^{-1}) at 1X field rate. The herbicide screenings were followed by dose-response assays of the most resistant ryegrass population at eight rates (0.5, 1, 2, 4, 8, 16, 32, and 64X), compared to a susceptible population at six rates (0.0625, 0.125, 0.25, 0.5, 1, and 2X). The initial screening, as well as dose-response experiments, were conducted in a completely randomized design with three replications and two experimental runs. Survivors were characterized as highly resistant (0-20% injury) or moderately resistant (21-79%) and susceptible (<80% injury). Results showed a high level of resistance to individual ALS- and ACCase-inhibitor herbicides and also cross and multiple herbicide resistance in four populations. Ratio of GR_{50} values indicated that most resistant population had 20-, 87-, and 161- fold resistance to pinoxaden, mesosulfuron, and pyroxsulam, respectively. GR_{50} ratio for diclofop-methly-resistant population could not be calculated as none of the doses provided 50% control. No resistance to glyphosate and pyroxasulfone was observed in these populations. As Italian ryegrass is an obligate outcrosser, multiple herbicide resistance can spread to neighboring fields though hybridization. The proliferation of multiple herbicide resistant ryegrass is a challenge to sustainable wheat production in Virginia and warrants Integrated Weed Management strategies.

PERSISTENCE AND RELOCATION OF DISLODGABLE HERBICIDE RESIDUE FROM SIMULATED RAINFALL FOLLOWING GLYPHOSATE TREATMENT TO DORMANT ZOYSIAGRASS TURF. Navdeep Godara*, Clebson G. Goncalves, Jordan M. Craft, Shawn Askew; Virginia Tech, Blacksburg, VA (27)

Turfgrass managers are concerned about sporadic injury to zoysiagrass from glyphosate during winter dormancy. The severity of glyphosate injury to zoysiagrass increases with increasing green leaves, but injury also occurs following glyphosate application to fully dormant zoysiagrass, especially when maximum-allowable glyphosate rates are applied prior to rapid post-dormancy shoot emergence. Glyphosate residue deposited on dormant zoysiagrass leaves could be dislodged and relocated to subtending green shoots during post-dormancy transition. Research experiments were conducted to determine how simulated raindrop quantity affects the relocation of glyphosate and colorant solution from treated dormant zoysiagrass leaves to underlying green shoots and to assess the response of post-dormant zoysiagrass to glyphosate dislodged and relocated via simulated rainfall, dew, or physical disturbance. Increasing the number of simulated raindrops per 0.5 cm^{-2} from 1 to 20 removed 95% of the applied colorant from treated dormant zoysiagrass leaves. Exposure of underlying filter paper was inversely related to loss from dormant zoysiagrass leaves. Maximum exposure of a single subtending green shoot was 5% of applied colorant and occurred at 5 simulated raindrops 0.5 cm^{-2}, which would approximate a rainfall of 3.35 mm. Furthermore, glyphosate applied to dormant zoysiagrass followed by 3 mm simulated rainfall or dew and wiping 7-9 days later, when turf had reached 15% green cover, reduced zoysiagrass clipping biomass by 35% and 72%, respectively. Glyphosate applied to dormant zoysiagrass can dislodge and injure newly developing green shoots during post-dormancy transition. Injury by dislodged and relocated glyphosate can be mitigated by irrigating at 1 day after glyphosate treatment to dormant turf or ensuring that the first irrigation or rainfall following herbicide treatment is at least 12 mm.

PHYTOENE DESATURASE INHIBITION WITH TOPICAL APPLICATIONS OF SIRNA.

Caroline E. Barrett*, Ramsey Lewis, Ralph Dewey, Ramon G. Leon; North Carolina State University, Raleigh, NC (28)

RNA interference, (RNAi) can silence the expression of genes blocking metabolic pathways. This technology can be used to block metabolism in new ways (i.e., new mechanisms of action) to control weeds. The present study explored the use of RNAi to block the production of phytoene desaturase (PDS) in Amaranthus palmeri S. Watson, Nicotiana benthamiana Domin, Nicotiana tobaccum L., and Arabidopsis thaliana (L.) Heynh. dsRNA solutions of various concentrations were applied to leaves and tissue bleaching was assessed to determine RNAi efficacy. Bleached tissue was quantified per square centimeter, with minimal differences between the different doses of dsRNA. However, there was a difference in efficacy between new and old plant tissues.

POSTEMERGENCE HERBICIDES APPLIED IN CONJUNCTION WITH NITROGEN FOR FALSE-GREEN KYLLINGA (KYLLINGA GRACILLIMA) CONTROL.

Katherine H. Diehl*, Trevor S. Watson, Matthew T. Elmore, Daniel P. Tuck; Rutgers University, New Brunswick, NJ (29)

False-green kyllinga (*Kyllinga gracillima*; FGK) is a rhizomatous sedge species that can be difficult to control with single herbicide applications. The objective of this research was to evaluate the effect of nitrogen fertilizer on the efficacy of three herbicides for FGK control. Field experiments were conducted in 2021 and 2022. The 2021 experiment was initiated on 15 July at the Rutgers Adelphia Research and Extension Farm in Freehold, NJ, on perennial ryegrass (*Lolium perenne*) infested with false-green kyllinga. The 2022 experiment was initiated on 30 June at the Rutgers Hort Farm No. 2 (HF2) in North Brunswick, NJ on a mixed stand of creeping bentgrass (*Agrostis stolonifera*) and tall fescue (*Festuca arundinacea*) infested with false-green kyllinga planted in 2018 from a biotype collected from a local golf course fairway. FGK cover was >70% when both experiments were initiated. Treatment factors consisted of single applications of three herbicides (granular pyrimisulfan at 50 g ha^{-1}; halosulfuron at 70 g ha^{-1} + NIS at 0.25% v/v; and sulfentrazone + carfentrazone applied at 560 + 30 g ha^{-1}) and four nitrogen rates (urea 46-0-0 applied at 0, 13, 25, or 50 kg N ha^{-1}) arranged in a complete factorial. Treatments were replicated four times and arranged in a RCBD with 0.9 by 2.0 m plots. Halosulfuron and sulfentrazone + carfentrazone were applied with a CO_2-powered single nozzle boom in 374 L ha^{-1} of water carrier though an AI9504EVS nozzle. Nitrogen and pyrimisulfan treatments were applied using a shaker jar and irrigated into the soil immediately after application. False-green kyllinga cover was evaluated visually on a 0 (i.e., no cover) to 100 (i.e., complete cover) percent scale, and grid intersect counts were conducted 12 weeks after treatment (WAT) and at the conclusion of the 2022 experiment at 52 WAT. Data were analyzed using the GLIMMIX procedure in SAS ($P = 0.05$) and Fisher's protected LSD test was used to separate means. Data for the 2021 experiment and 2022 experiment are presented separately as year-by-treatment interactions were detected.In 2021, nitrogen had no effect on herbicide efficacy ($P=0.07$) at 12 WAT, possibly due to excellent efficacy of all herbicide treatments. There was a non-significant tendency for FGK cover to decrease with N rate for pyrimisulfan from 21% when no N was applied to 9% at 12 WAT when N was applied at 50 kg ha^{-1}. Halosulfuron and sulfentrazone + carfentrazone were the most effective herbicide treatments and resulted in < 3% FGK cover across N rates. In the 2022 experiment, halosulfuron applied alone resulted in 74% FGK cover at 52 WAT while N applied at 13 kg ha^{-1} reduced FGK cover to 43%; N applied at 50 kg ha^{-1} further reduced cover to 10%. Nitrogen had no effect on FGK control provided by sulfentrazone + carfentrazone and pyrimisulfan at 52 WAT. FGK cover was 55 to 70% for pyrimisulfan treatments and 59 to 69% for sulfentrazone + carfentrazone treatments at 52 WAT. This research found applying nitrogen in conjunction with halosulfuron can increase false-green kyllinga control, but nitrogen effects were not consistent across year and location.

REDUCED ATRAZINE RATES FOR WEED CONTROL IN CORN.

Akashdeep Singh Brar*[1], Milos Viric[1], Mark J. VanGessel[2], Kurt M. Vollmer[3], Vijay Singh[1]; [1]Virginia Tech, Painter, VA, [2]University of Delaware, Georgetown, DE, [3]University of Maryland, Queenstown, MD (30)

The US Environmental Protection Agency (EPA) is proposing restrictions to atrazine in watersheds where elevated levels have been detected. During the recent atrazine re-registration process, the EPA proposed reducing the aquatic ecosystem level of concern (CE-LOC) from 15 ppb to 3.4 ppb. The CE-LOC is the 60-day average concentration of atrazine that, when exceeded, presents a greater than 50% chance of a negative effect on the community structure and function of an aquatic environment. The proposed label limits the maximum atrazine application rate to 2 lb./A per season for all areas, a half-pound lower than previously allowed. Most of the corn growing areas in Delmarva peninsula will be affected by these changes and this will significantly impact corn weed control programs. Two multi-state studies were conducted at three sites (one site each in Virginia, Delaware and Maryland) to determine how to maximize weed control with the proposed atrazine rates. Herbicide applications were applied at planting and included atrazine alone and in combination with S-metolachlor (Dual II Magnum), pyroxasulfone (Zidua) and/or mesotrione (Callisto) as well as a prepackaged mixtures consisting of bicyclopyrone, mesotrione, acetolchor, clopyralid, rimsulfuron, and S-metolachlor (example, Acuron Flexi, Lexar EZ, Resicore, Instigate, Corvus) etc. Studies conducted in Virginia indicated that reduced rate of atrazine (reduced rate; 20 fl oz/A) alone at preemergence does not provide excellent control of weeds, however, weed control was 100% when atrazine (20 fl oz/A) was tankmixed with S-metolachlor, pyroxasulfone or mesotrione. Another study where PRE herbicides were followed by postemergence (POST) treatments indicated that glyphosate + glufosinate (POST) treatments were marginally inferior (90%) to treatments that included atrazine and/or Halex GT and/or Acuron GT Flexi which provided season-long excellent control. The studies will expand knowledge of chemical strategies to help control weeds at reduced atrazine rates. This project addresses weed management in terms of effectiveness, resistance management, as well as its adaptability to the specific needs of the eastern shore region.

RESPONSE OF HERBICIDE-RESISTANT ANNUAL BLUEGRASS BIOTYPES TO ENDOTHALL.

Juan Romero*, John M. Peppers, Shawn Askew; Virginia Tech, Blacksburg, VA (31)

Endothall is a serine/ threonine inhibitor that was introduced in the 1950s for weed control in aquatic environments; later, it was registered for use in row crops and ornamental turf. Endothall has been utilized in cool-season turfgrass for control of annual bluegrass. However, its use was limited due to phytotoxicity and the appearance of more effective and affordable herbicides. Nonetheless, due to the increasing resistance of annual bluegrass to widely used herbicides, endothall represents a novel mode of action, and the hypothesis is that this herbicide will control annual bluegrass resistance biotypes in the same way it controls susceptible ones. For this project, the objective was to de determine the sensitivity of five known resistance annual bluegrass biotypes named (ALS1, ALS2, EPSP1, EPSP2, and MULTI) to glyphosate and trifloxysulfuron to six endothall doses (0, 500, 1000, 2000, 4000, 8000 g ai ha^{-1}). Two greenhouse studies were conducted at Glade Research Center in Blacksburg, VA. The study was designed as an RCBD with four replicants, and treatments included a 6 x 6 factorial arrangement with six levels of endothall rates applied to six different annual bluegrass biotypes. I_{50} and I_{90} control was visually assessed, and required rates were calculated via nonlinear regression. It is likely that changes in light intensity have influenced the results. Disparity between field experience and the first greenhouse study led the researchers to move the plants outside the greenhouse for the second study. At least 1.3 more endothall was needed to control AB in the first trial (T1) compared to the second one (T2). From the studied biotypes, only one, ESPS2, required more endothall (2.7X) compared to the susceptible to achieve I_{50} and I_{90} values. This biotype is currently under investigation for a potential nontarget resistance mechanism. This study suggests that endothall is able to control most, but not all, herbicide resistance biotypes of annual bluegrass.

CROP DIVERSITY AND COVER CROP MANAGEMENT IMPACTS ON WEED ABUNDANCE IN FORAGE GRAIN CROP SYSTEM.

Noelle A. Connors*[1], John M. Wallace[2], Heather Karsten[2]; [1]Pennsylvania State University, State College, PA, [2]Penn State University, University Park, PA (32)

ABSTRACT NOT AVAILABLE

SEED KILL OF PROBLEMATIC WEED SPECIES IN WHEAT AND SOYBEAN BY TWO SEED IMPACT MILLS.

Eli C. Russell*, Kevin Bamber, Michael L. Flessner; Virginia Tech, Blacksburg, VA (33)

Seed impact mills, like the Redekop Seed Control Unit (SCU) and the integrated Harrington Seed Destructor (iHSD), are aftermarket modifications that mount on the back of a combine. During harvest, any weed seeds that exit the combine in the chaff fraction will pass through the mill and they will be killed before they are spread back out into the field. The purpose of this research was to determine what the seed kill of problematic weed species in wheat and soybean of two different mills, the Redekop SCU and the iHSD. All testing was conducted using stationary test stands. Ten different species were tested that are problematic in soybeans and six different species were tested for wheat. Species tested were consistent across both mills. Two experimental runs with five replicates each were tested for each species. Results indicated that >97% of all weed seeds of the tested species in soybeans were killed by either mill. For problematic weeds like Palmer amaranth (*Amaranthus palmeri* S. Watson) and common ragweed (*Ambrosia artemisiifolia* L.), seed kill was >99.5% for both species in both mills. In wheat, seed kill was >99% for all species tested except for Italian ryegrass (*Lolium perenne* L. *ssp. multiflorum* (Lam.) Husnot). Italian ryegrass had the lowest reported seed kill at 91.4% and 92.7% for the iHSD and Redekop SCU, respectively, then the other species tested. However, these results are similar to other tests in Australia on a closely related species. These data indicate a promising outlook for seed impact mills. With high seed kill by the mills, there is potential to significantly reduce the number of weed seeds being returned to the soil seed bank during harvest.

STANDARDIZATION OF UNMANNED AERIAL SYSTEM-BASED HERBICIDE APPLICATION IN CORN.

Fatemeh Esmaeilbeiki*[1], Dhiraj Srivastava[2], Daniel Martin[3], Vijay Singh[1]; [1]Virginia Tech, Eastern Shore Agricultural Research And Extension Centre, VA, [2]Donald Danforth plant sciense center, St Louis, MO, [3]USDA, College Station, TX (34)

Unmanned aerial systems (UAS) have the potential for managing weeds with herbicide spray applications. Previously, UAS-based spray applications have indicated excellent weed control in corn, and soybean when applied at postemergence (POST), operating at 3 m altitude, and using 18.7 and 37.4 L ha^{-1} volume. In 2023, additional studies were conducted in corn at Virginia Tech-Eastern Shore Agricultural Research and Extension Center, Painter, VA, to evaluate the efficacy of UAS-based spray applications for preemergence (PRE) herbicide treatments. Two experiments were conducted, 1) evaluating three volumes, and 2) evaluating two nozzle types (air-induction and regular flat fan). Grass and broadleaf weed seeds were broadcasted at corn planting and weed density was recorded before spray application. For experiment 1, a spray mixture was applied at 18.7 L ha^{-1} (UAS-2; 2 gallons/A), 37.4 L ha^{-1} (UAS-4; 4 gallons/A), 74.8 L ha-1 (UAS-8; 8 gallons/A), and compared with a CO_2 pressurized backpack sprayer at 140 L ha^{-1} (BP-15). Artificial samplers; and water-sensitive papers (WSP) were placed in plots (50 x 20 m) to determine droplet spectra, and visual weed control was recorded. These experiments were conducted in a completely randomized block design with four replications using S-metolachlor and a mixture of atrazine, bicyclopyrone, mesotrione, and *S*-metolachlor. Analysis of Variance (ANOVA) was performed and Tukey's HSD (a=0.05) was employed for means separation. Spray application treatments showed significant differences in weed control w.r.t 18.7 L ha^{-1} and 37.4 L ha^{-1}, however, both 37.4 L ha^{-1} and 74.8 L ha^{-1} volume provided similar weed control. Experiment 2 emphasized the importance of nozzle selection, indicating that air-induction nozzles effectively reduce drift and postemergence weed control but may not be optimal for preemergence applications, which require finer droplet sizes. Regular flat fan nozzle provided 30% greater weed control than air-induction nozzle when used at the preemergence stage. Results have indicated that current UAS technology can be used for both PRE and POST herbicides but more studies and technological improvements are required for standardization.

STATEWIDE SCREEN OF NORTH CAROLINA ITALIAN RYEGRASS (LOLIUM MULTIFLORUM) POPULATIONS WITH RESIDUAL HERBICIDES.

Diego J. Contreras*, Jackson W. Alsdorf, Ronel J. Argueta, Edgar Posadas, Colden Bradshaw, Wesley Everman; North Carolina State University, Raleigh, NC (35)

Blue River Technology and John Deere's See & Spray™ technology is a computer vision and machine learning system adapted for sprayers that provides the ability to selectively target weeds within a crop. The See & Spray™ ATM machine is a condensed version of the commercial See & Spray™ Ultimate, which allows for small plot research in weed science. See & Spray™ ATM operators can adjust several settings, one of the options provides the ability to select a sprayer sensitivity setting. Sprayer sensitivity is defined as the confidence level of the sprayer to trigger a spray on a targeted weed. With a lower sensitivity, the sprayer will need a high confidence level to spray a weed, with a higher sensitivity, the sprayer will require a lower confidence level to spray the same weed. While lower sensitivity settings may potentially reduce sprayer output, more weeds may be missed when compared to a higher sensitivity setting. Weed size is generally thought of as an important factor when adjusting sprayer sensitivity settings. This experiment was deisgned with the objective of determining what factors influence See & Spray™ sprayer sensitivity selection in cotton and soybean models and what adjustments should be made to reduce missed weeds. The factors of interest were sprayer travel speed, weed species, weed size and weed location within the crop row. Two experiments were established consisting of two experimental runs in both cotton and soybean crops. Weeds were transplanted into plots, consisting of grass weeds only (Texas panicum and large crabgrass), broadleaf weeds only (Palmer amaranth and sicklepod) and mixed weeds which contained the species previously mentioned. Weeds were smaller than two inches and larger than four inches, and transplanted in the furrow and in the bed besides the crop row. Data logs were collected with the See & Spray™ ATM machine by simulating sprays at travel speeds of 8, 12 and 15 miles per hour and using high, mid and low sprayer sensitivity settings. The data logs processed to determine a percentage of targeted weeds that would have triggered a spray. Data were subjected to an ANOVA using PROC GLIMMIX in SAS 9.4 to determine statistical differences between factors. In both cotton and soybean, sprayer sensitivity setting was the greatest factor impacting targeting weeds. Travel speed was not a significant factor in targeted weeds. Weed size and weed species impact targeted weeds within a sprayer sensitivity setting. Targeted weeds at a low sprayer sensitivity setting could not be increased by adjusting other parameters listed.

SURVEY OF HERBICIDE RESISTANT PALMER AMARANTH POPULATIONS IN NORTH CAROLINA.

Jackson W. Alsdorf*, Diego J. Contreras, Ronel J. Argueta, Colden Bradshaw, Edgar Posadas, Wesley Everman; North Carolina State University, Raleigh, NC (36)

Amaranthus palmeri is a troublesome weed commonly found in multiple row crops in North Carolina. Many cases of herbicide resistance have been documented across the state, and a handful of statewide screens have been done to determine the distribution of these resistance cases. In the fall of 2022, North Carolina was grid sampled in an effort to collect Palmer amaranth seed heads out of fields with observed populations. The objective of this study is to understand the frequency and distribution of resistance to multiple herbicides. 137 populations were collected and seed heads were threshed then stored in a freezer until the screen. Seeds were planted in flats in a greenhouse, and each population consisted of five treatments: untreated check, fomesafen (281 g ai ha), glufosinate (596 g ai ha), dicamba (562 g ae ha), and 2,4-D (798 g ae ha). Ratings were taken weekly on a scale of 0-100 for each individual plant, with 0 being no control, and 100 being plant death. The 28 DAT rating was used to average the total control for all plants across each population within each treatment. If control was 50% or less overall, then that population was determined to be resistant to that specific herbicide. While no populations were found to be considered resistant to glufosinate, dicamba, or 2,4-D, seven populations were found to have potential resistance to fomesafen. These results are helpful tools for growers and county agents as they put together effective weed control programs.

USING UV-FLUORESCENT DYE TO MEASURE MULTIPASS DEPOSITION PATTERNS OF AERIAL AND GROUND APPLICATION EQUIPMENT.

Daewon Koo*, Navdeep Godara, Juan Romero, Shawn Askew; Virginia Tech, Blacksburg, VA (37)

Uniform spray deposition is crucial for achieving desirable outcomes in broadcast pesticide application. Previous methods for assessing spray deposition patterns have had limited capabilities due to restricted sampling areas and low sampling resolution within the field. A technique utilizing a fluorescence dye as a proxy for the pesticide, coupled with aerial image analysis, has been developed. Studies were conducted to evaluate the spray deposition patterns of a multipass spray drone equipped with flat-fan nozzles or air-induction nozzles, a simulated ride-on sprayer, and a spray gun with a flooding nozzle. A fluorescent dye solution (Cartax DP Liquid, Heubach Colorant USA LLC, Charlotte, NC) was applied at a rate of 3.7 L per hectare, regardless of the application method. Six ultraviolet lamps uniformly illuminated a quarter of each plot, and nighttime aerial images were captured. Each image was overlaid with a 5000-point grid at a resolution of 0.27 dm^2 using FieldAnalyzer (Green Research Services LLC, Fayetteville, AR) to calculate the average hue within the image. The average hue was then converted to the estimated fluorescence dye dosage based on a logarithmic standard curve that related four reference plots treated with known dye concentrations. To capture each 5.5 m x 9.7 m plot, four images were collected, each having their own 4-dose reference plots for standard comparison. Additional samples of each reference plot were collected on 10-cm diameter Petri dishes and dosage was further confirmed via fluorescence spectrometry. Proportions within the target area that were less than 50% and greater than 150% of the targeted rate were considered under and over-application, respectively. The proportion of grid positions with over 50% of the target rate on non-target perimeter areas was calculated as drift. This novel method was considered successful based on the agreement between fluorescence spectrometry and digital image analysis results. Across 48 samples at each of four standard doses, the average deviation was 20% and means were within 7% of hypothetical rates with the highest accuracy at the 1X rate. Errors are assumed to be associated with nonuniformity in standard dose delivery as these were delivered with a hand-held spray boom. The results suggested that the multipass spray drone application, regardless of nozzle type, had 24% more overapplication than the ride-on sprayer. Both spray drone applications exhibited at least 35% more drift than ground applications. Chem gun applications had minimal drift but suffered from over and under-application.

WEED EMERGENCE TIMING IN A WARMER WORLD.

Nasib Koirala*[1], Carolyn Lowry[2], Richard G. Smith[3], Alexandra Contosta[3]; [1]The Pennsylvania State University, State College, PA, [2]The Pennsylvania State University, University Park, PA, [3]University of New Hampshire, Durham, NH (38)

Warmer temperatures driven by global climate change are expected to impact weed emergence dynamics. To examine how these elevated temperatures will alter the timing of weed emergence, a study was conducted in 2022-23 in Rocksprings, PA, with in-situ temperature modification. Open Top Chambers (OTCs), constructed from fiberglass panels, were used to increase temperature in the experimental plots. The emergence of six weed species: *Chenopodium album*, *Ambrosia artemisiifolia*, *Amaranthus retroflexus*, *Abutilon theophrasti*, *Setaria faberi*, and *Digitaria sanguinalis* was counted from April to August of 2023 in plots with or without OTCs. The average monthly increase in soil temperature at 3 cm depth with the OTCs was 0.54°C. Reduction in total emergence was observed in the warming treatments for all species except in *Abutilon theophrasti*. This might be due to increased seed mortality in higher temperatures for other species while velvetleaf was protected due to its thicker seed coat. *Abutilon theophrasti* and *Ambrosia artemisiifolia* tended to emerge early in the season compared to the other weed species studied. Even though total emergence was changed, the pattern for emergence was similar under warming temperatures for individual weed species. The results indicate that weeds with better seed physical defense might be more competitive in future warmer climatic conditions.

WEED IDENTIFICATION USING IMAGE ANALYSIS AND MACHINE LEARNING.

Akhilesh Sharma*, Vipan Kumar, Louis Longchamps; Cornell University, Ithaca, NY (39)

The rapid evolution of herbicide-resistant (HR) weed populations and increasing cost of their control warrant the development of site and specie-specific weed management strategies (SSSWMS). Successful deployment of various SSSWMS further requires rapid and accurate detection of weed species in various cropping situations. The main objectives of this research were to (1) develop an annotated image database of cocklebur, dandelion, common waterhemp, Palmer amaranth and common lambsquarters to fine-tune object detection algorithms (YOLOv8 and Detectron2 (Faster RCNN)), and (2) investigate the comparative performance (speed and accuracy) of both algorithms in detecting those weed species. Field sites with natural infestations of cocklebur, dandelion, and common lambsquarters were identified during 2023 growing season at Cornell University Musgrave Research Farm, Aurora, NY. In addition, a grower field in Seneca County, NY was identified with natural infestations of common waterhemp in soybeans. Images of each weed species were manually collected at early growth stages (cotyledons to 30 cm tall) using a GoPro Hero RGB camera. Palmer amaranth images (cotyledons to up to 76 cm tall) were collected from corn and soybean fields during 2022 growing season at Kansas State University Agricultural Research Center near Hays, KS. All collected images were pre-processed to a size of 640 x 640 pixels, augmented by blurring, adding noise, and rotations. Total number of images for each weed species were as following: 858 for Palmer amaranth, 720 for common lambsquarters: 352 for common waterhemp, 227 for dandelion, and 191 for cocklebur.Images were then manually annotated with bounding boxes, producing 2,348 annotated images. The annotated images were used to train the Detectron2 and YOLOv8 algorithms. Mean average precision (mAP) and inference speed metrices were used to evaluate the accuracy and speed of both algorithms in identifying and locating each weed species. Results indicated that YOLOv8 outperformed Detectron2 in detecting all weed species. The YOLOv8m algorithm achieved an mAP score of 93% @0.5 IoU and an inference speed of 23 milliseconds with a Tesla T4 GPU. In contrast, the Detectron2 with Fast R-CNN configuration had an mAP of 68% @0.5 IoU and an inference time of 210 milliseconds. The mAP @0.5 IoU with YOLOv8 testing was 90% for Palmer amaranth, 87% for common lambsquarters,93% for common waterhemp, 97% for dandelion, and 96% for cocklebur. These preliminary results conclude that the YOLOv8m algorithm can play a crucial role for weed detection more accurately and about 9 times faster than Detectron2 in agricultural fields. Contact E-mail: as3774@cornell.edu

CORN YIELD LOSS PREDICTION WITH UAV MEASURED CROP-WEED LEAF COVER.

Avi S. Goldsmith*, Robert Austin, Charlie W. Cahoon, Ramon G. Leon; North Carolina State University, Raleigh, NC (40)

The Kropff and Spitters model predicts yield loss based on weed and crop leaf area ratios. When the model was proposed, there was no practical way to measure leaf area index in real time during commercial production. Remote sensing technologies such as UAV's may make data collection for the model significantly more efficient. The goal of this study was to validate the Kropff and Spitters model with UAV aerial imagery of crop and weed leaf area cover (Lwc) in maize (Zea mays L.), increasing data collection efficiency. Through Lwc and a weed competitive ability coefficient (i.e., q), the model predicts yield loss based on the relative size and emergence time of the weeds compared to the crop. Aerial images of maize and weed leaf cover from four locations in North Carolina were analyzed with supervised object-based classification. A correction factor (i.e., c) was developed to compensate for competition differences due to leaf are index vs. leaf cover. This allowed for prediction of maize yield loss with adequate accuracy in a fast non-destructive manner. The integration of aerial image collection prior to canopy closure and mathematical modeling may allow growers to anticipate weed-driven yield losses and make the necessary management and financial decisions to increase weed control and profits.

CONFIRMATION OF AN ATRAZINE RESISTANT PALMER AMARANTH POPULATION IN NORTH CAROLINA.

Ronel J. Argueta*; North Carolina State University, Raleigh, NC (41)

Atrazine has historically been efficacious herbicide in North Carolina, where almost every hectare of corn receives a pre- and postemergence application. In 2016, a farmer in Washington Co. North Carolina reported a control failure on *Amaranthus palmeri* S. Watson (Palmer amaranth) in a corn field treated with atrazine. The objectives were to evaluate the response of the putative atrazine-resistant population from Washington Co. (R) at different atrazine rates and determine effective herbicides to control the putative atrazine resistant population. Two atrazine-susceptible populations from Edgecombe Co. (S) and Johnston Co. (S) were used susceptible controls The experimental design was completely randomized with four replications. Atrazine was applied at 0, 56, 177, 560, 1770, 5600 and 17,700 g ai ha^{-1} and included crop oil concentration (1% v v^{-1}). Treatments were applied to plants 7.6 to 9 cm in height. Plant survival was evaluated 21 days after treatment. The LD_{50} for each population were calculated from a three-parameter log-logistic model. A resistance ratio was calculated for each atrazine-susceptible population. The LD_{50} value was 1354 g ai ha^{-1} for R population. The R/S were 6.8 to 8.7 compared to the S populations. The Washington Co. population has evolved atrazine resistance. Subsequently, these *A. palmeri* populations were treated with 7 herbicide modes of action for further characterization under similar conditions and evaluations. All populations exhibited resistance to thifensulfuron and glyphosate with survival =70% and =30%, respectively. 2,4-D and dicamba provided variable control on all populations. The population is resistant to thifensulfuron, atrazine and glyphosate. Keywords: atrazine; glyphosate; herbicide resistance; Palmer amaranth; *Amaranthus palmeri*; thifensulfuron; weed management

BEYOND BIOMASS: WEED SUPPRESSION BENEFITS OF CEREAL RYE AT HIGH SEEDING RATES.

Laurel Wellman*[1], John M. Wallace[2]; [1]The Pennsylvania State University, State College, PA, [2]Penn State University, University Park, PA (42)

Organic no-till soybean (*Glycine max* L.) production relies on roll-crimped cereal rye (*Secale cereale* L.) mulch as the primary means of weed suppression. Yet achieving an adequate mulch remains challenging in crop rotations where rye sowing is delayed following corn (*Zea mays* L.) harvest. To investigate the effects of cultural management practices on cereal rye productivity for weed suppression, we established a field study in the fall of 2021 and 2022. Using a two-factor, randomized complete block design with three replicates, we tested the effects of rye sowing arrangement (grid sown vs. typical 19cm rows) and four cereal rye seeding rates with an unseeded control (0, 31, 63, 126, 188 kg ha^{-1}) on rye productivity (spring groundcover and biomass) and weed suppression (weed biomass) during the subsequent soybean crop. Higher rye seeding rates increased groundcover in one year and while there was no main effect of sowing arrangement on groundcover, there was a seeding rate by sowing arrangement interaction in one year, with row sown rye achieving higher groundcover than grid sown rye. Seeding rate moderately increased rye biomass production in both years. Notably, biomass production was low (< 6000 kg ha^{-1}) in both years due to delayed rye sowing and low soil fertility following corn harvest. Yet, weed biomass was significantly reduced when sowing rye at high rates. Structural equation modeling revealed that while both groundcover and rye biomass increased with seeding rate, neither service mediated the negative relationship between seeding rate and weed biomass, suggesting a weed suppression mechanism from cereal rye other than biomass and groundcover contributes to improved weed suppression at high rye seeding rates. Our research suggests that although high seeding rates cannot achieve the groundcover or biomass potential of early sown cereal rye, there may be a weed suppression benefit to utilizing high rye seeding rates in cover crop-based no-till soybean production.

INFLUENCE OF LASER INTENSITY ON SMOOTH CRABGRASS CONTROL IN CREEPING BENTGRASS PUTTING GREENS.

Juan Romero*, Shawn Askew; Virginia Tech, Blacksburg, VA (43)

The increase of herbicide-resistant weeds and regulatory pressure are limiting effective weed control options in turfgrass. Organic methods for weed control in professional turf management lack selectivity and are expensive. By targeting weed control treatments to individual plants, selectivity is improved and chemical costs are reduced, but such approaches have been cost prohibitive due to human labor requirements. Recent advances in artificial intelligence may speed development of robotic solutions for turfgrass weed management. Autonomous systems can treat individual weeds at a fraction of the costs of human labor. Past experiments conducted mainly in agricultural systems show that laser beam treatments are more effective when applied to seedlings. In these systems, laser power and spot size have determined weed control success, but turfgrass systems differ in that more mature plants can't be killed by rapid laser exposure to seedling stems. Research is needed to determine the influence of laser intensity on weed control in turf. Studies were conducted using a 10-W laser engraver to treat digitally selected weed objects with 1-mm spaced lines at 250, 500, 750, and 1000 mm/min laser speed and 1, 2, or 3 laser passes. Digital photos were collected at 7, 14, and 21 DAT using a custom-made lightbox, which also supported the laser. Binary images of polygon-selected smooth crabgrass objects were dilated 40 pixels and treated with 1-mm spaced laser lines at 100% power and treatment-associated speeds and number of passes in a creeping bentgrass putting green. Due to frost-associated smooth crabgrass decline, only creeping bentgrass recovery could be assessed as influenced by laser intensity. Speed was positively correlated to creeping bentgrass recovery when assessed as percentage green pixels detected in the treated area compared to pretreatment values based on digital image analysis using hue and saturation thresholds. Two passes decreased creeping bentgrass recovery 13% compared to one pass, but did not differ from three passes. Generally, laser intensity was inversely related to creeping bentgrass recovery. Future research will explore more variables like pattern, spacing, weed size, and weed species to determine optimal laser intensity for weed control in managed turfgrass.

WEED CONTROL AND WATERMELON YIELD RESPONSE TO COVER-CROP TERMINATION TIMING.

Carrie Mansue*[1], Thierry E. Besancon[1], Kurt M. Vollmer[2]; [1]Rutgers University, Chatsworth, NJ, [2]University of Maryland, Queenstown, MD (44)

Weeds can significantly reduce yields in plasticulture production systems. Weeds growing between rows can compete with the crop, interfere with harvest, and serve as hosts for pathogens and other plant pests. Weeds are managed in these systems mechanically, through cultivation, by mowing, and/or by hand-weeding. However, these tactics often need to be repeated to control weeds that germinate and emerge throughout the growing season. Cover crops have been evaluated as an alternative weed management practice in row crops; however, limited research has been conducted in vegetables, particularly those grown on plastic. The main objective of this project was to evaluate weed control and watermelon yield in response to oat cover crop planted between plasticulture rows. The trial was conducted as a split plot design with four replications. Main plot was termination with paraquat (560 g ha^{-1}) at either boot or heading stage of spring oat whereas subplot included either spring oat cover crop between rows or bare ground. Trials were conducted at the Wye Education and Research Center in Queenstown, MD, and at the Rutgers Agricultural Research and Extension Center in Bridgeton, NJ. two separate locations, New Jersey and Marylan Data collected included cover crop biomass at termination, % weed control, % weed coverage and weed counts in 0.25m^2 quadrats immediately prior to termination, 7 days after termination (DAT), 21 DAT, and 35 DAT. as well as weed biomass 115 days after cover crop seeding. Watermelons were harvested over three times with fruits counted, weighed, and graded according to USDA grade and standard guidelines. Overall and across locations, oat termination at boot or heading stage had no significant effect on weed biomass, cover crop biomass and average watermelon marketable weight. Weed counts taken 35 DAT, and averaged over termination timing, indicated that an oat cover crop reduced weed density 47% as compared to bare ground. Weed biomass collected 115 days after seeding was 75% lower in row middles cover-cropped with spring oat as compared to bare ground. Total watermelon yield and fruit count were not influenced by the presence of the oat cover crop nor termination stage. However, the proportion of marketable fruits was significantly higher in spring oat cover crop plots (62%) compared to bare ground plots (44%). Mean weight of individual marketable fruits was 6.1 kg where spring oat cover crop was seeded between rows as compared to 4.7 kg in the absence of cover crop. Results suggest that integrating cover crops into vegetable production systems can be effective for suppressing weeds and improving yields. Use of a spring seeded oat cover crop between rows in plasticulture production systems demonstrated significant potential for reducing weed competition and improving watermelon marketable yield while reducing the number of herbicide applications required for control of competing weeds.

CONFIRMATION OF GLYPHOSATE RESISTANT PALMER AMARANTH (AMARANTHUS PALMERI) BIOTYPES IN NEW YORK AND RESPONSES TO ALTERNATIVE CHEMISTRIES.

Aleah L. Butler-Jones*[1], Elizabeth C. Maloney[2], Sarah Morran[3], Todd A. Gaines[3], Lynn M. Sosnoskie[2]; [1]Cornell University, Ithaca, NY, [2]Cornell University, Geneva, NY, [3]Colorado State University, Fort Collins, CO (45)

Palmer amaranth (*Amaranthus palmeri* S. Wats.) is one of the most troublesome weeds in North America due to its rapid growth rate, substantial seed production, competitiveness, and the evolution of herbicide resistant biotypes. Though frequently encountered in the US South and Midwest, Palmer amaranth was recently identified in soybean fields in Genesee, Orange, and Steuben Counties in NY, where glyphosate was the primary herbicide for in-crop weed control. This research, conducted in 2022 and 2023, aimed to 1) describe the dose response of three putative resistant NY Palmer amaranth biotypes to glyphosate, 2) determine their mechanisms of resistance, and 3) assess their sensitivity to other POST herbicides commonly used in NY crop production. Based on the effective dose necessary to kill 50% of plants (ED_{50}), the NY Palmer amaranth biotypes were 42- to 67-times more resistant to glyphosate compared to the glyphosate-susceptible (GS) biotype tested. Additionally, the NY biotypes had elevated *EPSPS* gene copy numbers, ranging from 25 to 185, which are harbored within the extrachromosomal circular DNA (eccDNA) replicon. The NY biotypes were poorly controlled by label rates of chlorimuron-ethyl and chloransulam-methyl. Some variability was observed among biotypes in response to atrazine, and hydroxyphenylpyruvate dioxygenase (HPPD)-inhibiting herbicides. All NY biotypes were effectively controlled by label rate applications of herbicides within WSSA Groups 4, 7, 10, 14, and 22. Additional research is needed to confirm whether the NY Palmer amaranth biotypes display multiple resistance to other groups of herbicides and develop alternative management strategies suitable for NY crop production.

SEED IMPACT MILLS' WEED SEED KILL RATE AND HORSEPOWER DRAW AS INFLUENCED BY CHAFF FLOW RATE AND MOISTURE.

Eli C. Russell*, Kevin Bamber, Michael L. Flessner; Virginia Tech, Blacksburg, VA (46)

Combine modifications for harvest weed seed control, like the Redekop Seed Control Unit (SCU) and the integrated Harrington Seed Destructor (iHSD), may be affected by changes in crop yield and harvest residues' moisture as they can have an impact on chaff flow rate and chaff moisture, respectively. The purpose of this research was to determine how varying chaff flow rates and chaff moisture content in wheat and soybean chaff would affect the seed kill rate and horsepower draw of two different impact mills, the Redekop SCU and the iHSD. Four different chaff flow rates were tested for wheat and soybean, 0.75, 1.5, 2.25, and 3.0 kg sec^{-1}, which spans approximately 0.5 to 2-fold a combine's capacity. Additionally, four chaff moisture contents were tested for wheat and soybean, 12, 18, 24, and 30%. All testing was conducted using stationary test stands. Three weed species were tested for wheat, Italian ryegrass (*Lolium perenne* L. *ssp. multiflorum* (Lam.) Husnot), canola (*Brassica napus* L.), and hairy vetch (*Vicia villosa* Roth), and four weed species were tested for soybean, Palmer amaranth (*Amaranthus palmeri* S. Watson), common ragweed (*Ambrosia artemisiifolia* L.), barnyardgrass (*Echinochloa crus-galli* (L.) P. Beauv.), and morningglory spp. (*Ipomoea* spp.). The data indicate that horsepower draw for both mills increases in wheat and soybean as the chaff flow rate increases. However, as chaff moisture increases only the iHSD had an increase in horsepower draw in wheat. The effects of chaff flow rate and chaff moisture on seed kill varied depending on the mill and weed species. Italian ryegrass, Palmer amaranth, and common ragweed most commonly had decreases in seed kill in response to increases in chaff flow rate or chaff moisture across both mills. While there was a decrease in seed kill, Palmer amaranth and common ragweed experienced at most a 1% decrease in seed kill across the tested values for either chaff flow or chaff moisture. Italian ryegrass experienced the largest decrease in seed kill in response to an increase in chaff flow rate. Seed kill dropped by 7.5% across the tested values for the Redekop SCU and by 18.4% for the iHSD. Despite the increase in horsepower draw and the decrease in seed kill rate, the data indicates the potential for seed impact mills to operate in less-than-ideal conditions while still providing high kill rates in soybean (>98%) and wheat (>74%).

EVALUATING HERBICIDE TOLERANCE ON COSMOS BIPINNATUS.

Jose H. de Sanctis[1], Brock A. Dean[1], Jacob C. Forehand[2], James H. Lee*[1], Zachary R. Taylor[3], Charlie W. Cahoon[1]; [1]North Carolina State University, Raleigh, NC, [2]North Carolina State Unviersity, Raleigh, NC, [3]North Carolina State University, Sanford, NC (47)

Evaluating POST Herbicide Options for *Cosmos bipinnatus* J.H. Lee, C.W. Cahoon, Z.R. Taylor, B.A. Dean, J.C. Forehand J.S. De Sanctis During the summer of 2023, herbicide safety screening was conducted on *Cosmos bipinnatus* currently being planted by the N.C. Department of Transportation (NC DOT) Roadside Environmental Unit for inclusion in the wildflower program. *C. bipinnatus is* a summer annual that is planted extensively across the state. Two locations were planted, one at Central Crops Research Station in Clayton, NC on July 25th, 2023. The second site was provided by Garrett Seed Farm in Four Oaks, NC on September 3, 2023. Both locations had pendimethalin at 1064 g ai/ha and s-metolachlor at 1092 g ai/ha applied at planting. The following herbicides were applied (rates in parenthesis)postemgergence: fluometuron (1212 g ai/ha), fluometuron (2241 g ai/ha), 2,4-DB (175 g ae/ha), 2,4-DB (210 g ae/ha), 2,4-DB (245 g ae/ha), a premix of flufenacet + metribuzin (378 g ai/ha), a premix of flufenacet + metribuzin (474 g ai/ha), tolpyralate (29 g ai/ha), topramezone (12 g ai/ha), flumetsulam (1 g ai/ha), flumetsulam (6 g ai/ha), fluroxypyr (118 g ai/ha), tembotrione (92 g ai/ha), acifluorfen (140 g ai/ha), imazapyr (70 g ai/ha), imazapyr (350 g ai/ha), pyroxasulfone (64 g ai/ha), sulfentrazone (140 g ai/ha), isoxaflutole (53 g ai/ha), simazine (1121 g ai/ha), halauxifen (5 g ai/ha), tiafenacil (25 g ai/ha), fluridone (168 g ai/ha), flumioxazin (420 g ai/ha), a premix of flumioxazin + prodiamine (980g ai/ha). When required by label recommendations, crop oil concentrate, methylated seed oil, and non-ionic surfactant were added at recommended rates. In this trial fluometuron at 1212 g ai/ha acts as a comparison treatment for C. bipinnatus, as this is the current labeled standard practice used by the NCDOT. Clayton was treated on August 10th, 2023, and Four Oaks was treated on September 18, 2023. At treatment, the average height of C. bipinnatus was 15 cm. Treated plots were 3 meters wide by 9 meters long. At each location, running checks were left between plots. Treatments were replicated 4 times throughout each location. Treatments were applied using a CO2 pressurized backpack sprayer equipped with Tee jet flat-fan AIXR 11002 nozzles, calibrated to deliver 140 liters per hectare. Visual ratings of general phytotoxicity were recorded at one, two, and three weeks after treatment. Heights were taken four weeks after treatment. Percent bloom reduction was also visually estimated at peak bloom. It was found that fluometuron (1212 g ai/ha), fluometuron (2241 g ai/ha), pyroxasulfone (64 g ai/ha), a premix of flufenacet + metribuzin (378 g ai/ha), a premix of flufenacet + metribuzin (474 g ai/ha), isoxaflutole (53 g ai/ha), flumioxazin (420 g ai/ha), and a premix of flumioxazin + prodiamine (980g ai/ha) caused little to no reduction in bloom compared to the running check. There appear to be several viable herbicide options that could be used over the top of C. bipinnatus. Another year of testing is needed to determine if these herbicides are safe. After further testing, 24(c) labels can be pursued so that these products may be incorporated into weed management plans by the NCDOT wildflower program

INVESTIGATING THE MECHANISM OF PARAQUAT RESISTANCE IN ITALIAN RYEGRASS POPULATIONS FROM NORTH CAROLINA.

Jose H. de Sanctis*[1], Charlie W. Cahoon[1], Travis W. Gannon[1], Wesley Everman[1], Katherine M. Jennings[1], Zachary R. Taylor[2]; [1]North Carolina State University, Raleigh, NC, [2]North Carolina State University, Sanford, NC (48)

Italian ryegrass (*Lolium perenne* L. ssp *multiflorum* (Lam.) Husnot) is a winter annual weed species commonly found in row crops and non-crop areas across North Carolina. Populations resistant to herbicide groups 1, 2, and 9 have already been confirmed in the state. Consequently, paraquat has become a common burndown option for Italian ryegrass control. In 2021, paraquat-resistant populations were confirmed in the Southern Piedmont region of the state. The objective of this study was to investigate the mechanism of resistance from the recently found paraquat-resistant Italian ryegrass populations from North Carolina through absorption, translocation, and metabolism studies with ^{14}C-paraquat material. Two separate studies were conducted to investigate the mechanism of paraquat resistance. Both studies were conducted as a randomized complete block design with five replications, two runs, two resistant (R1 and R2), and two susceptible (S1 and S2) populations. In the absorption and translocation study, plants were harvested 0, 4, 12, 24, 48, 96, and, 192 hours after exposure to ^{14}C-paraquat and dissected in four samples (above treated leaf, below treated leaf, treated leaf, and roots). Samples were then dried until constant weight, combusted in a biological oxidizer, and radioactivity was determined by liquid scintillation spectrometry. The amount absorbed was converted to a percentage of the total radioactivity applied and translocation as a percentage of absorbed. In the metabolism study, plants were harvested at 0, 24, 48, 96, and, 192 hours after treatment, samples were then processed and analyzed via HPLC-DAD spectrometry. There were no differences in the total paraquat absorbed nor the rate of paraquat absorption between resistant and susceptible populations with total amount of paraquat absorbed ranging from 89-94%. Furthermore, susceptible populations translocated nearly 4 times more ^{14}C-paraquat out of treated leaf when compared to resistant populations. There was no indication of paraquat metabolism, and all populations had similar amounts of parent material. These results indicate that restricted paraquat movement is associated with paraquat resistance and further investigation is needed to determine the exact resistance mechanism.

CHARACTERIZING VARIOUS SPRAYER SETTINGS ON JOHN DEERE SEE & SPRAY™.

Diego J. Contreras*[1], Lauren M. Lazaro[2], William L. Patzoldt[2], Jackson W. Alsdorf[1], Ronel J. Argueta[1], Edgar Posadas[1], Wesley Everman[1]; [1]North Carolina State University, Raleigh, NC, [2]Blue River Technology, Sunnyvale, CA (49)

Italian ryegrass (Lolium multiflorum) is one of the most troublesome weeds for wheat production throughout Southern US (Webster 2008). The inability to control this weed can result in reduced yields, reduced quality, or both (Grey 2012). Additionally, herbicide resistance is a persistent problem throughout the United States, with over 36 reported cases. (Heap 2023). In North Carolina, effective post-emergence control options for L. multiflorum have become very limited. ACCase and ALS inhibitor herbicides, commonly used post-emergence in wheat, have lost efficacy due to an increase in herbicide resistance. Weed control programs for L. multiflorum rely on pre-emergence herbicides due to the limited post-emergence options. Two of the most commonly used are flumioxazin and pyroxasulfone. However, there have been reported L. multiflorum escapes with these pre-emergence options. It is imperative to understand the distribution of surviving populations around the state. To understand the current distriubution of control failures, an experiment was conducted to screen a statewide collection of L. multiflorum populalations with two commonly used residual herbicides in wheat, flumioxazin and pyroxasulfone. 118 populations across the state of North Carolina were collected from wheat fields in 2022. A dose response assay was conducted with three known herbicide susceptible populations collected in 2012 to determine a discriminating dose at the LD90 level for each flumioxazin and pyroxasulfone. The 118 populations were screened by treating the populations with the discriminating dose for each herbicide. At the discriminating dose (21 grams a.i. ha-1) Flumioxazin failed to control (<50%) 6 populations, and 45 other populations exhibited poor control (51 – 69%). Pyroxasulfone failed to control (<50%) 4 populations, and 49 other populations exhibited poor control (51 – 69%). Populations with control failures were spread across the state. The large number of L. multiflorum populations from the 2022 collection that exhibit poor control or lower could be a premature indication of increasing herbicide resistance cases around North Carolina.

POLLINATORS ALTER THEIR FORAGING BEHAVIOR DEPENDING ON HERBICIDE MODE OF ACTION AND PLACEMENT ON WHITE CLOVER.

Navdeep Godara*, Shawn Askew; Virginia Tech, Blacksburg, VA (50)

The recent decline in pollinator abundance threatens global food production. Pollinators risk exposure to insecticide residue when visiting weedy flowers in urban landscapes. Several herbicide modes of action (MOAs) control white clover, but the effect on pollinator foraging behavior remains unaddressed in scientific literature. Research experiments were conducted at Blacksburg, VA, in 2023 to assess the effect of herbicide MOAs on honey bee foraging visitation, white clover floral density, and white clover floral quality. Treatments included a nontreated control, Trimec Classic® (2,4-D + MCPP + dicamba); Halo 75 WDG® (halosulfuron); Dismiss® (sulfentrazone); and Pylex® (topramezone). A subsequent study evaluated how synthetic auxin placement on white clover flowers affects honey bee visitation. 2,4-D + MCPP + + dicamba eliminated honey bee visitation completely at 1 d after treatment (DAT), but the same impact on bee visitation required 14 d following topramezone treatment. Halosulfuron and sulfentrazone transiently affected honey bee visitation, as they reduced <80% of honey bee foraging visits. Honey bee visitation on topramezone-treated flowers was least dependent on flower quality when compared to other herbicides. Synthetic auxins reduced white clover density to 100% at 14 DAT, but other treatments reduced floral density to <65%. Additionally, synthetic auxin placement on white clover flowers is essential to achieve rapid deterrent activity on pollinator visitation. Honey bee foraging behavior is temporally dependent on herbicide MOAs. Our research provides practitioners with transient, food-resource preserving or long-term pollinator foraging deterrents to minimize pollinator exposure to insect-pest-management treatments. Future research will assess the temporal response of nectar production and in situ bee exposure to pesticide and tracer dye following deterrent treatments.

EXPLORING ANAEROBIC SOIL DISINFESTATION AS NEW NON-CHEMICAL TACTIC FOR WEED MANAGEMENT IN ORGANIC WATERMELON PRODUCTION.

Sohaib Chattha*, Matthew A. Cutulle; Clemson University, Charleston, SC (51)

Soilborne pathogens and weeds are limiting factors in organic watermelon (*Citrullus lanatus*) production. Yellow nutsedge (*Cyperus esculentus* L.) and palmer amaranth (*Amaranthus palmeri* S. Watts) are two major problematic weeds in Southeastern U.S. organic vegetable production systems. Inefficient non-chemical tactics are an impediment to curtail diseases and weeds in organic watermelon; and necessitate the adoption of alternative pest management strategies. Anaerobic soil disinfestation (ASD) has the potential to fit into current pest management practices. ASD is a potential alternative to chemical fumigation which is incorporation of labile carbon (C) sources into the soil, followed by tarping the soil with plastic mulch, and irrigating the soil to the saturation. Changes in soil microbial communities and production of volatile organic compounds during anaerobic decomposition are the main mechanisms that are believed to kill and/or deactivate the soilborne pathogens and weeds seeds. ASD has been shown to improve crop vigor and weed control in various specialty crops. However, the effectiveness of ASD has not been proven in watermelon. A field study was conducted at the Clemson University Coastal Research and Education Center in Charleston, SC to specifically look at the impact of chicken manure + molasses (CM+M), and cotton seed meal (CSM) on grafted vs non-grafted watermelon yield. The study was designed in a randomized complete block design. The treatments were assigned as ASD with powerhouse non-grafted and powerhouse scion grafted on Carolina strongback rootstock. All treatments that went anaerobic removed the iron oxide paint from indicator of reduction in soils (IRIS) tubes and reduced yellow nutsedge counts when compared to the treatment that did not go anaerobic. At the time of watermelon harvesting, total number of yellow nutsedge counts were recorded as 65, 25, and 22 in control, CSM, and CM+M, respectively. Based on weed control and yield assessments, using CM+M to facilitate ASD is an ideal practice for growing organic watermelon in South Carolina.

SCREENING STRATEGIES AND HERITABILITY ESTIMATION FOR ALLELOPATHY IN RYE (SECALE CEREALE).

Aliyah Warris*[1], James B. Holland[2], Virginia M. Moore[3], S Chris Reberg-Horton[1], Steven B. Mirsky[4], Ramon G. Leon[1]; [1]North Carolina State University, Raleigh, NC, [2]USDA ARS, North Carolina State University, Raleigh, NC, [3]Cornell University, Ithaca, NY, [4]USDA ARS, Beltsville, MD (52)

The demand for cereal rye as a cover crop is increasing in organic and conventional farming systems. The benefits of rye as a cover crop include increased water infiltration, reduced soil erosion, and the production of large amounts of residues that add organic matter to the soil. In addition to benefits to the soil, rye as a cover crop can suppress weed germination and establishment by forming a physical barrier on the soil surface and also via allelopathy. Allelopathy is the production and release of chemicals from root exudates that injure seedling tissues, which results in reduced emergence and growth of surrounding weeds. Rye breeding programs for cover cropping have focused on biomass production, but little attention has been paid to breeding for increased allelopathy. In this study, 38 crosses derived from lines previously identified as highly allelopathic accessions and commercial rye varieties 'ND Gardner' and 'Aroostook' were screened for allelopathic activity using the *in vitro*, equal-compartment-agar method and lettuce as the bioindicator. After *in vitro* phenotyping, rye seedlings were transplanted to pots filled with loam soil to assess allelopathic activity in soil under greenhouse conditions. Lettuce growth suppression was used as a surrogate of weed suppression, and vegetative growth data was collected, including rye dry biomass (g), tiller number, leaf area (cm^2), and plant height (cm). Genetic correlations between level of allelopathy and vegetative growth of rye individuals were determined. Results showed 14 of the lines demonstrated high allelopathic activity compared to the rye-free control (> 50% growth reduction). Heritability estimates (H^2) *in vitro* were 0.61 for the ND Gardner crosses, 0 for the Aroostook crosses, and 0.47 for the ND Gardner and Aroostook data combined. In soil, H^2 was 0.0435 for the ND Gardner lines, 0 for the Aroostook lines, and 0.40 for the ND Gardner and Aroostook data combined. The H^2 estimates indicate that allelopathy has a genetic component large enough to rationalize breeding for increased allelopathy in cereal rye.

UAV SPRAY HEIGHT INFLUENCES DROPLET VAPORIZATION AND WEED CONTROL EFFICACY.

Daewon Koo*, Navdeep Godara, Juan Romero, John M. Peppers, Shawn Askew; Virginia Tech, Blacksburg, VA (53)

Recent research evaluating spray pattern deposition from agricultural spray drones indicated that total spray deposition was inversely related to drone height. It was hypothesized that droplet evaporation contributed to losses in field-measured deposition since the spray drone utilizes a spraying system that generates very fine droplets and operating heights ranged from 2 to 10 m. Additional research was conducted to measure actual evaporative losses from droplets in the range of sizes indicative of the drone's spray system and to determine if smooth crabgrass control in a turfgrass system was negatively impacted by spray height. Our first objective was to determine how droplet diameter influences evaporative loss of discretely sized droplets delivered at two distances separated by 5-m travel through the air in an indoor, controlled environment. Our second objective was to determine the influence of spray height on smooth crabgrass control with quinclorac herbicide in managed turfgrass. The laboratory study was conducted at the Glade Road Research Facility and utilized a custom droplet generator consisting of a 2.5-cm disc spinning at 3750 rpm to deliver discrete droplets through a narrow aperture. Droplets consisted of water and Blazon Spray Pattern Indicator at 1:1 v/v and were measured based on digital analysis of stains on craft paper positioned at 0.5 and 5.5 m below the droplet generator. Droplets of 106, 150, 178, 216, and 271 µm were separated by horizontal distances from the generator and were confirmed by capturing droplets in a dimethylpolysiloxane, biphasic oil solution; photographing; and digitally measuring droplet diameters. Likewise, stain objects on paper were related to droplet size based on previous studies using oil-captured versus paper stain size relationships. Results indicate that 106 µm droplets completely evaporated while 150 to 271 µm droplets lost 37 ± 2 % of their initial diameter after traveling 5 m through the air. Using reported droplet spectra for the TeeJet XR Flat Fan 11001 nozzle and estimating evaporative loss based on laboratory studies, total spray loss at a 5 m height was estimated at 53%. To evaluate the effects of this evaporative loss on the field performance of herbicides, managed turfgrass infested with approximately 30% smooth crabgrass was treated with a DJI MG-1P spray drone at 28 L ha^{-1} to apply quinclorac at 0.84 kg ai ha^{-1} at 2, 6, and 10 m spray heights. A ground sprayer comparison was included that consisted of the same herbicide delivery but at 373 L ha^{-1} using drift reduction nozzles. Smooth crabgrass control 28 d after treatment was inversely related to height with applications made at 10 m controlling smooth crabgrass 15% compared to 70% control when quinclorac was applied at the 2-m height. These data suggest that agricultural spray drones should be operated at minimum-allowable heights to prevent pesticide product loss due to evaporation.

HIGHER SEEDING RATES FOR WEED SUPPRESSION IN ORGANIC KERNZA (INTERMEDIATE WHEATGRASS) PRODUCTION.

Natasha Djuric*[1], Antonio DiTommaso[2], Jacob Jungers[3], Tim Crews[4], Jose Franco[5], Leonardo Deiss[6], Steve Culman[7], Matthew Ryan[8]; [1]Cornell University, Soil and Crop Sciences, Ithaca, NY, [2]Soil and Crop Sciences Section, School of Integrative Plant Science, Cornell University, Ithaca, NY, [3]University of Minnesota, Twin Cities, MN, [4]The Land Institute, Salina, KS, [5]Dairy Forage Research Center, Madison, WI, [6]The Ohio State University, Wooster, OH, [7]Washington State University, Pullman, WA, [8]Cornell University, Ithaca, NY (54)

Kernza® (*Thinopyrum intermedium*) is the first commercialized perennial grain and can be planted once and harvested for several years. Despite demonstrated ecosystem services and potential for dual-use forage and grain production, Kernza adoption has been limited due to low grain yields. Weed-crop competition is a major production constraint, especially in organic production during the first year. Increasing crop seeding rate is a cultural management practice that can reduce weed establishment and improve crop competitiveness. However, intraspecific competition can also reduce Kernza grain yields, particularly after the first grain harvest. Research on the tradeoff between weed suppression and intraspecific competition in Kernza is lacking and could inform management recommendations. A field experiment was established in late summer 2022 in New York (two locations), Ohio, Wisconsin, Minnesota, and Kansas to quantify the effects of seeding rate on weed suppression and Kernza yield. At each location, Kernza (cv. 'MN-Clearwater') was planted at five seeding rates (6, 17, 28, 39, and 50 kg pure live seed/ha) and data were collected on species-level weed biomass and the density, ground cover, biomass and grain yield of Kernza. Results indicate that planting Kernza at higher seeding rates markedly reduced weed biomass during the establishment year without affecting grain yield. Future research should explore the effects of high seeding rates on weed suppression and grain yields of Kernza during the second and third year after planting, and determine the economic optimum seeding rate based on seed costs, weed suppression, and grain yield over the life of the Kernza stand.

IMPACT OF PARTIAL SALTWATER AGROECOSYSTEMS ON WEED COMPETITION IN WATERMELON.

Matthew A. Cutulle[1], Brian Ward[2], Joseph F. Bazzle*[2], Sandra Branham[3], Amnon Levi[4], Hailey Schopp[2]; [1]Clemson University, Charleston, SC, [2]Clemson University, Clemson, SC, [3]Clemson, Charleston, SC, [4]USDA-USVL, Charleston, SC (55)

The issue of soil salinity as a major cause of poor soil health and crop yield loss has been of growing concern as climate change contributes to its effects. The objective of this research was to study the impact of increasingly saline soils on the relationship between grafted watermelons and yellow nutsedge, one of the major weeds in watermelon plasticulture. The seedless watermelon cultivar Melody was grown in a field after being grafted onto the C. maxima hybrid Carnivor and the C. amarus cultivar Carolina Strongback in addition to both a self-grafted and ungrafted control. The field was divided into four rows, which were irrigated with 0, 10%, 20%, and 30% dilutions of sea water for the duration of the experiment. A weed count was performed after one month and three months of irrigation. This demonstrated that salt had a significant effect on the total weed count at high concentrations, however the weeds demonstrated a much greater resistance to salt treatment than the watermelons in this trial. Based on this data, it is possible that salt intrusion events can contribute to increased weed related yield loss in watermelon crops.

GOOSEGRASS (ELEUSINE INDICA) ECOTYPES AFFECTED BY CULTURAL MANAGMENT AND PLANT GROWTH REGULATORS IN TURFGRASS.

Katherine H. Diehl*, Matthew T. Elmore; Rutgers University, New Brunswick, NJ (56)

Competitiveness of different goosegrass (*Eleusine indica*) ecotypes has not been investigated under turfgrass management regimens. The objective of this research was to conduct a two-year long field experiment to determine effects of mowing height, nitrogen fertilizer, and a plant growth regulator (PGR) on the competitiveness of two different goosegrass ecotypes in a field experiment. Research was conducted on Kentucky bluegrass (*Poa pratensis*), annual bluegrass (*Poa annua*) turf at Rutgers Hort Farm 2 research farm (North Brunswick, NJ). The site was mowed weekly to 4.0 cm starting 3 May 2023 and fertilized twice with slow-release fertilizer (25 kg N ha^{-1}) before the start of the experiment. Treatments consisted of three factors; 1) PGR and Nitrogen (N) fertilizer program (hereafter referred to as cultural management) 2) goosegrass ecotype and 3) mowing height. Treatments were arranged in 1 by 2 m plots as a strip-plot RCBD, with whole-plot factors of cultural management (non-treated, N only, PGR only, and N+PGR) and goosegrass morphologically distinct ecotypes (turf-type and ag-type) arranged as a 4-by-2 complete factorial. These treatments were replicated four times and strips of mowing height (2.0 cm and 4.0 cm) were imposed. Mowing height treatments were initiated 16 May 2023 and plots were mowed twice weekly until mid-October. Nitrogen treatments (urea 46-0-0; 25 kg N ha^{-1}) and PGR treatments (trinexapac-ethyl; 190 g ai ha^{-1}) were applied on the same day every 21 d from 19 May through 4 September using standard small plot spray equipment. N treatments were applied first and irrigated in immediately after application; PGR treatments were not irrigated. On 5 July 2023, 20 leaf-stage seedlings from each ecotype were transplanted into field plots in two rows on 13 cm inter-row spacing. Goosegrass tillers on each plant were counted one and three wks after transplanting (WAT). The number of surviving goosegrass plants in each plot was counted 2, 4, 6, 8 and 10 WAT. Goosegrass cover was evaluated visually and determined by grid intersect at 6 and 8 WAT. On 28 September, at 10 WAT, total aboveground goosegrass biomass was harvested. Data were analyzed in SAS (v 9.4) as a strip-plot RCBD and Fisher's protected LSD test (?=0.05) to separate means. Interactions between cultural management program (N and PGR treatments) and ecotype were detected in goosegrass cover and biomass data on each date. Mowing height had no affect on goosegrass biomass or cover. In plots not treated with N or PGR, the turf ecotype had more biomass than the ag ecotype (47 g and 6 g, respectively). N alone reduced goosegrass cover and biomass of the turf ecotype by >50% compared to the non-treated, but treatment did not affect the ag ecotype. PGR treatment increased biomass of both the ag and turf ecotypes by 31% and 549%, respectively, compared to the non-treated, and the turf ecotype treated with PGR alone had the greatest biomass of all treatments (63 g). N+PGR treatment reduced goosegrass biomass by 23% compared to the nontreated for the turf ecotype, but increased biomass of the ag ecotype to 234% of the nontreated. This research shows the PGR trinexapac-ethyl can increase the competitiveness goosegrass in turf, despite having a positive effect on turfgrass density. Nitrogen treatment reduced goosegrass competitiveness overall but varied by ecotype. Turf ecotypes were more competitive than ag ecotypes in turf. Trinexapac-ethyl increased the competitiveness of ag ecotypes more than turf ecotypes. A second run of this experiment will be conducted in summer 2024.

SURVEY OF HERBICIDE RESISTANT ITALIAN RYEGRASS POPULATIONS IN NORTH CAROLINA.

Jackson W. Alsdorf*, Diego J. Contreras, Ronel J. Argueta, Colden Bradshaw, Edgar Posadas, Wesley Everman; North Carolina State University, Raleigh, NC (57)

Lolium perenne L. ssp. multiflorum is a troublesome weed commonly found in North Carolina. In June of 2022, 118 samples were collected from around the state by taking mature weed seed heads from wheat fields. The objective of this study is to understand the frequency and distribution of resistance to multiple herbicides. Samples were threshed then stored in a freezer. In the greenhouse, seeds were germinated in flats, then transplanted into individual conetainers with 4 replications treated with one of 6 treatments: untreated check, glyphosate (1349 g ai ha), glufosinate (596 g ai ha), mesosulfuron (15 g ai ha), pinoxaden (61 g ai ha), and paraquat (562 g ai ha). Ratings were taken weekly on a scale of 0-100 for each individual plant, with 0 being no control, and 100 being plant death. Populations with 50% or less control with a specific herbicide are considered to be potentially resistant. 114 populations were found to be resistant to mesosulfuron, 17 to pinoxaden, 2 to glyphosate, 1 to paraquat, and 0 to glufosinate. One five-way resistant population was found, along with 10 three-way resistant populations. These results are helpful tools for growers and county agents as they put together effective weed control programs.

ESTABLISHMENT AND WINTER SURVIVAL OF JOHNSONGRASS (SORGHUM HALEPENSE) IN CENTRAL NEW YORK STATE.

Ryan P. O'Briant*[1], Lynn M. Sosnoskie[2], Antonio DiTommaso[3]; [1]Cornell University - Soil and Crop Sciences, Brooktondale, NY, [2]Cornell University, Geneva, NY, [3]Soil and Crop Sciences Section, School of Integrative Plant Science, Cornell University, Ithaca, NY (58)

ABSTRACT NOT AVAILABLE

USING 3-D IMAGING TO MAP BIOMASS DISTRIBUTION IN COVER CROPS.

April M. Dobbs*[1], Avi S. Goldsmith[1], Daniel J. Ginn[2], Søren K. Skovsen[3], Muthukumar V. Bagavathiannan[2], Steven B. Mirsky[4], Chris S. Reberg-Horton[1], Ramon G. Leon[1]; [1]North Carolina State University, Raleigh, NC, [2]Texas A&M University, College Station, TX, [3]Aarhus University, Aarhus, Denmark, [4]USDA ARS, Beltsville, MD (59)

Cover crops are an important tool for weed control, and biomass production is a key predictor of weed suppression ability. However, variability in biomass within fields poses a challenge for growers when estimating biomass production. Moreover, most available methods for estimating biomass are laborious and impractical on a field scale. This study used ground-based Structure-from-Motion (SfM) to estimate biomass in cereal rye (*Secale cereale* L.) and model biomass distribution using geostatistics. SfM was used to generate 3-D point clouds of cereal rye from videos collected with a GoPro camera over five fields in North Carolina during the 2022-2023 season. Biomass and crop height were measured in quadrats throughout each field. A model was generated to predict biomass based on point cloud pixel density and crop height. Measured biomass at termination was linearly related to pixel density multiplied by crop height (R^2 = 0.813) through levels of 9,000 kg ha^{-1}. Based on independent data validation, predicted biomass and measured biomass were linearly related (R^2 = 0.713). In areas of higher biomass, the model overpredicted by as much as 3,140 kg ha^{-1}, and in areas of lower biomass, it underpredicted by as much as 2,837 kg ha^{-1}. However, the spatial distribution of high and low biomass areas were the same for predicted and measured biomass. This method provides a robust prediction of rye biomass and can potentially be used by growers to nondestructively estimate biomass production and identify areas in the field that may be at greater risk of late-season weeds.

PLANTING TIME INFLUENCES ANNUAL FLOWER STRIP ESTABLISHMENT WHEN WEED ABUNDANCE IS LOW.

Sophie Westbrook*[1], Scott S. Morris[2], Rebecca Stup[1], Rosa Xia[2], Ryleigh Coffey[2], Antonio DiTommaso[2]; [1]Cornell University, Ithaca, NY, [2]Soil and Crop Sciences Section, School of Integrative Plant Science, Cornell University, Ithaca, NY (60)

Flower strips are an effective way to enhance agroecosystem biodiversity and ecosystem services. Most flower strips are composed of perennial species. Despite their ecological benefits, perennial flower strips are not widely adopted. Barriers to adoption include the long-term commitment required and concerns about weeds. This study explores whether annual flower strips might be accessible to more farmers. We conducted an on-farm experiment on five commercial farms in New York, USA. On each farm, we established four treatments. At maize planting time, we seeded an early-established planting (EP) treatment with a commercial mix of annual flowers. An early-established control (EC) was set up at the same time. Four weeks later, we prepared a new seedbed for late-established planting (LP) and late-established control (LC) treatments. We observed a significant effect of treatment on plant species richness and Shannon diversity ($P < 0.001$). Planted treatments were more diverse than control treatments. However, there was no effect of planting time (EP vs. LP) on diversity. Treatment also affected the number of flowering dicot species ($P = 0.001$), which was numerically highest in the EP treatment. Flowering species richness was positively associated with spider abundance in sweep-net samples. Overall, these results demonstrate that annual flower strip establishment is possible even under weedy conditions. In addition, they show that a delay in planting date does not eliminate the benefits of this practice. This information could help farmers make informed, site-specific decisions about whether flower strips are a good fit for their farms.

IMPACT OF ROOT SEGMENT LENGTH AND PLANTING DEPTH ON VEGETATIVE PROPAGATION OF COMMON MILKWEED (ASCLEPIAS SYRIACA).

Rebecca Stup*[1], Sophie Westbrook[1], Antonio DiTommaso[2]; [1]Cornell University, Ithaca, NY, [2]Soil and Crop Sciences Section, School of Integrative Plant Science, Cornell University, Ithaca, NY (61)

Asclepias syriaca L. (common milkweed) is both injurious to crops and capable of providing a range of ecological benefits, the most well-known being its function as a host plant for Danaus plexippus L. (monarch butterfly) larvae. As such, there is interest in both managing it as an agricultural weed and in propagating it for conservation purposes. The aim of this study was to determine the effect of root segment length and planting depth on total emergence and biomass of common milkweed propagated from lateral root buds. Root segments of 3, 8, or 15 cm were planted at depths of 3, 8, or 15 cm in a greenhouse trial arranged in a completely randomized design. At 10 weeks after planting, total emergence was noted, and the plants were harvested to assess above- and belowground dry biomass. Neither root segment length nor planting depth had an impact on total emergence after 10 weeks. Planting depth did not have an impact on above- or belowground biomass. Root segment length had a significant impact on biomass. Plants grown from 15 cm segments were larger than plants grown from 3 or 8 cm segments both above- ($p < 0.0001$) and belowground ($p = 0.0001397$). These results demonstrate that plowing to a depth of 15 cm is not sufficient to control established milkweed stands, as the root segments are not buried deeply enough to prevent reemergence. Those interested in propagating milkweed for conservation purposes may want to use root segments greater than 15 cm in length.

AN INTEGRATED APPROACH TOWARDS WEEDS MANAGEMENT.

Haroon Ur Rashid*; Dubai Community Management L.L.C (Dubai Holding), Dubai, United Arab Emirates (62)

A two-year study was conducted on integrated weed management in maize under different tillage regimes at Agricultural Research Station Swabi Khyber Pakhtunkhwa, Pakistan during Spring 2014 and subsequently repeated in 2015. The experiment was laid out at silt loam soil in Randomized complete block design (RCBD) with a split plot arrangement having three replications. Tillage regimes ((minimum, conventional and deep tillage) were kept in main plots (Factor A) and allelopathic plant residues (sorghum, sunflower and parthenium) as surface mulched in various combinations and their water extracts @ 15L each + atrazine @ ¼th of recommended dose were assigned to the sub-plots (Factor B), for weed management in maize. Data during both years (2014 and 2015) were recorded and analyzed for dry biomass of weeds 30 DAS, kernels ear-1, ear length (cm), kernel yield (kg ha-1) and cost benefit ratio. Foliar application of Sorghum + parthenium water extracts at 15 L integrated with a quarter recommended dose of atrazine (pre emergence) under conventional tillage regimes suppressed total weed dry biomass by 34 and 42% at 30 DAS during 2014 and 2015, respectively which increased maize kernel yield by 52 % over the weedy check and was almost equivalent to the label dose of atrazine (0.50 kg a.i ha-1) and also had the highest CBR (1:20.4). Among the soil mulch treatments, Sorghum + sunflower +parthenium each at 4 Mg ha-1 under deep tillage regimes suppressed weed dry biomass by 69 and 75 % at 30 DAS during 2014 and 2015 respectively, pooled data of both years (2014 and 2015) indicated increase in maize kernel yield by 54 % over control with CBR (1:15:2). However, the mulch treatments and the cost of deep tillage were uneconomical. Hand weeding under deep tillage regimes increased maize kernel yield by 46 % as compared to the weedy check. Based on current studies, it is concluded that foliar application of Sorghum + parthenium aqueous extract integrated with reduced atrazine dose are economical and eco-friendly having the highest CBR. Consequently, reliance on atrazine could be reduced by 75% resulting in environmental safety and sustainability, however further studies are suggested to fine tune our findings. Key words: Allelopathy, Integrated weed management, tillage, sustainable weed management, maize.

PARTHENIUM HYSTEROPHORUS L. IN THE AMERICAS: PREDICTING SUITABLE HABITATS BEYOND ITS NATIVE RANGE UNDER FUTURE CLIMATIC CONDITIONS.

Sarah E. Kezar*[1], Asad Shabbir[2], Antonio DiTommaso[3]; [1]Cornell University, Ithaca, NY, [2]Biosecurity & Food Safety, Department of Regional NSW, New South Wales, Australia, [3]Soil and Crop Sciences Section, School of Integrative Plant Science, Cornell University, Ithaca, NY (63)

Parthenium weed (*Parthenium hysterophorus* L.) is an annual weed of global significance, beyond its native range from the southern region of North America to the northern parts of South America. The consequences of climate extremes have already been widely observed and predicted to alter the distribution of invasive populations in Australia, Asia, Africa, and Europe. However, this threat is yet to be understood for native populations in North American habitats and cropping systems. The development of a CLIMEX distribution model considered global habitat suitability data to predict the distribution of parthenium weed in North America for current and future (+3 °C) climates. Secondly, irrigation was added to the climate change scenario (+3 °C) to predict how extra moisture may affect its distribution. Parthenium weed is currently found in parts of the southern portion of the Cotton Belt, while increases in habitat suitability in a future climate will likely increase its range and stand densities. Similarly, the northern region of the Cotton Belt is projected to become more climatically suitable under climate change. The range of habitat suitability may further extend to more northern latitudes that encompass the southern Great Plains, go as far north as Virginia along the east coast, and across the west coast. When the irrigation scenario was added to the CLIMEX program, more parts of the semi-arid regions from West Texas to southern California became suitable due to irrigation producing extra moisture to support the establishment and growth of parthenium weed. Transient populations of parthenium weed have been reported in more northern latitudes, but the distribution of this species in both current and future climates is likely limited by cold winter temperatures.

EFFECTS OF CROP ROTATION ON COMMON WATERHEMP POPULATION DYNAMICS: AN INVESTIGATION USING A PERIODIC MATRIX MODEL.

Huong T. X. Nguyen*[1], Matt Liebman[2]; [1]Cornell University, Ithaca, NY, [2]Iowa State University, Ames, IA (64)

Crop biophysical characteristics and management practices can affect weed population dynamics in various ways. To track changes a common waterhemp (*Amaranthus tuberculatus* (Moq.) J.D. Sauer) population from seed in the soil seedbank to new seed deposition to the soil seedbank, we used a chain of six periodic matrices in each of nine crop environments crossed with two corn weed management regimes to project population trajectories in two scenarios of plant fecundity, representing two levels of weed control efficacy (high and low). The chain of periodic matrices described the response to crop management activities of the population of interest throughout a calendar year. To parameterize the model, we used values derived from both scientific literature and our own field observations. Each crop environment identified a crop species (corn, soybean, oat, or alfalfa) in a rotation (2-year, 3-year, or 4-year). The crop sequences in the 2-year, 3-year, and 4-year rotations were, respectively, corn - soybean; corn - soybean - oat intercropped with red clover; and corn - soybean - oat intercropped with alfalfa - alfalfa. When waterhemp control efficacy was high, the 3-year rotation appeared to be the most reliable in depleting the soil seedbank. When waterhemp control efficacy was low, the 4-year rotation appeared to be the least risky for preventing waterhemp outbreaks. The slower rates of population growth in the more diverse rotation were attributed to lower population growth rate (?) in the oat, red clover, and alfalfa crop environments (cool-season crops). In addition to population projection, we determined thresholds for mature plant density in the three rotations for stabilizing population size using the parameter inputs from the low control efficacy scenario. The differences in mature plant density thresholds were more pronounced between the 2-year and 4-year rotations than any other pairwise comparison.

RECENT ACTIVITIES OF THE MASSACHUSETTS INVASIVE PLANT ADVISORY GROUP, 2022 - 2023.

Randall G. Prostak*; University of Massachusetts, Amherst, MA (65)

RECENT ACTIVITIES OF THE MASSACHUSETTS INVASIVE PLANTS ADVISORY GROUP (MIPAG). R. Prostak*, University of Massachusetts - Amherst, Amherst, MA ABSTRACT Formed as an outgrowth of the Massachusetts Native Plant Advisory Group in 1995, the Massachusetts Invasive Plant Advisory Group (MIPAG) is a voluntary collaborative representing organizations and professionals concerned with the conservation of the Massachusetts landscape. MIPAG members represent research institutions, non-profit organizations, the green industry, and state and federal agencies. Because of this diversity of support, the findings and recommendations of the group will encourage a cooperative effort among every organization, agency, and citizen concerned with the threat of invasive plants. MIPAG with the assistance and direction of Les Mehrhoff of the University of Connecticut developed and adopted definitions, created a list of plant to be evaluated, and develop a set of biologically based criteria upon which to evaluate objectively plants suspected to be invasive in Massachusetts. Initially, a total of 85 species were evaluated and identified as Invasive, Likely Invasive, or Potentially Invasive. Species evaluated in which sufficient information or evidence was lacking for an adequate evaluation were designated as plants not meeting criteria (do not list at this time). MIPAG worked with the Massachusetts Department of Agricultural Resources to develop and implement the Massachusetts Prohibited Plant List, https://www.mass.gov/info-details/massachusetts-prohibited-plant-list. In the last two years, MIPAG has evaluated a few new species suspects of being invasive in Massachusetts. Recent additions to the list of invasive plants include scotch broom (*Cytisus scoparius*), weeping love grass (*Eragrostis curvula*), Japanese black pine (*Pinus thunbergii*), European alder or black alder (*Alnus glutinosa*) and callery pear (*Pyrus calleryana*). Changes in MIPAG's invasive plant evaluation criteria where updated based on recommendations of Bethany Bradley at the University of Massachusetts and Regional Invasive Species and Climate Change (RISCC). Their research on the impact on range-shifting due to climate charge on invasive plants was incorporated in order to be develop a more proactive stance on potential new invasive plants that might be problematic in the region. Background information about the Massachusetts Invasive Plant Advisory Group can be found at https://massnrc.org/MIPAG/ MIPAG during the initial evaluation process anticipated the possibility that sterling cultivars of plants determine to be invasive may be developed and seek to find a place in the ornamental nursery market. The decision to address this issue was made to leave room for discussion if and when this should occur. In the last several years, plant breeders have attempted to bring "sterile" or "low-sterility" cultivars to the market. MIPAG realize that it does not have the expertise to develop an evaluation protocol for these cultivars the group reached out to a few ornamental plant breeders for assistance. MIPAG has been working closely with Thomas Ranney, North Carolina State University and Ryan Contreras, Oregon State University. Thomas Ranney and Ryan Contreras have received an AmericanHort Foundation's Horticultural Research Institute (HRI) grant to support the development of an evaluation protocol for "sterile" or "low- sterility" ornamental cultivars.

A NOVEL PREMIX OF QUIZALOFOP-P-ETHYL AND GLUFOSINATE-AMMONIUM DEMONSTRATING BROAD-SPECTRUM WEED CONTROL.

Daniel L. Kunkel*; AMVAC, Plainsboro, NJ (66)

A unique formulation, a soluble liquid formulation created through ProLease™ Technology (containing 27.6 g/L of quizalofop-P ethyl plus 280.4 g/L of glufosinate-ammonium) has been developed for postemergence nonselective control of emerged grass and broadleaf weeds in glufosinate-tolerant crops including: canola (*Brassica napus*), cotton (*Gossypium herbaceum*), soybean (*Glycine max*) and for glufosinate- and quizalofop-P-ethyl-tolerant corn (*Zea mays*). Other crops will include orchard floor applications to the pome fruit (crop group 11-10) and stone fruit (crop group 12-12). The label registration is currently pending for these crops. The premix use rates will be from 496 to 968 g/ha for broadleaf weeds up to 8 cm, up to 13 cm tall for many grass weeds and up to 76cm for (non-Enlist®) volunteer corn, Johnsongrass (*Sorghum halepense) and* Grain Sorghum *(Sorghum bicolor)*. The premix has demonstrated excellent control of weeds, equal to or greater than a tank-mix of commercial formulations of quizalofop-P ethyl and glufosinate-ammonium applied at equivalent active ingredient rates. The premix did not influence broadleaf weed control as compared to glufosinate applied alone and grass weed control increased compared to quizalofop-p ethyl applied alone. The premix was an excellent choice when applied after soil-residual herbicides or tank-mixed with a Group 15 herbicide, and a sequential application of the premix resulted in greatest control. Optimum timing for weed control was observed when applied early to mid-postemergence with a spray volume containing medium to course textured spray quality with petroleum oil concentrate plus ammonium sulfate adjuvants. The premix does not provide residual weed control. The premix may be mixed with other registered pesticides, but antagonism of grass control may occur with some products, particularly herbicides in Group 2, 4, 6, and 14. The new pre-mix showed excellent crop tolerance with less than 10% for all crops and applications. This premix is the only commercial premix containing a grass herbicide such as quizalofop with a postemergence broadleaf herbicides and the only premix containing glufosinate-ammonium.

DEVELOPMENT OF CONVINTRO™ BRAND HERBICIDES FOR MANAGING AMARANTHUS SPECIES IN CORN AND SOYBEAN: FIELD PERFORMANCE UPDATE.

John Buol*, Carl Coburn, Richard Leitz, Zewei Miao, Emily Scholting; Bayer, St. Louis, MO (67)

The continued development and spread of herbicide resistance constitutes a major threat to the efficiency and profitability of corn and soybean production. Weeds such as some Amaranthus species have developed resistance to multiple herbicide modes- and sites- of action and are among the most challenging broadleaf weeds in North America. Bayer CropScience is developing a herbicide technology that features the use of diflufenican, a herbicide from a new site of action for control of Amaranthus spp in corn and soybean production systems in North America, pending registration with the U.S. EPA and Canada PMRA. Given the increasing challenge of managing herbicide-resistant weeds, diflufenican is being evaluated in field trials in North America for residual activity on Amaranthus spp. and crop selectivity in soybean and corn. A preliminary update on diflufenican development will be given featuring performance data from field trials. Pending registration with the U.S. EPA and Canada PMRA, diflufenican would enable a new weed management tool that should be used in combination with other weed management practices as part of an integrated weed management plan.

FROM LIBERTY 280 (RACEMIC GLUFOSINATE AMMONIUM) TO LIBERTY ULTRA (L-GLUFOSINATE AMMONIUM) HERBICIDE, POWERED BY GLU-L TECHNOLOGY.

Liam J. Vincent[1], Shawn C. Beam*[2], Eric Schultz[1], Alice Harris[1], Marcel Kienle[3], Ryan Aldridge[1], Sam Willingham[1], Ingo Meiners[1], Siyuan Tan[4]; [1]BASF, Rtp, NC, [2]BASF, Boalsburg, PA, [3]BASF, Limburgerhof, Germany, [4]BASF, Cary, Nc 27519, Usa, NC (68)

Glufosinate ammonium has been utilized as a postemergence herbicide in glufosinate tolerant cropping systems for nearly 30 years. To this point, all glufosinate herbicides registered for use in the United States have been in the form of a racemic mixture, including Liberty® 280 herbicide from BASF. Racemic mixtures of glufosinate contain a 1:1 ratio of D-glufosinate and L-glufosinate enantiomers. The L-isomer of glufosinate has herbicidal activity while the D-isomer has negligible herbicidal activity as it does not inhibit glutamine synthetase (GS) the target enzyme. For years it has been known that the two enantiomers have existed together in racemic mixtures; however, a resolved isomer form of L-glufosinate ammonium has never been commercialized in the United States. Pending registration, BASF intends to launch Liberty® ULTRA Herbicide, Powered by Glu-L Technology in 2024. Liberty ULTRA herbicide is the resolved isomer of L-glufosinate ammonium, and Glu-L Technology is the patent protected manufacturing process by which the D-isomer of glufosinate is enzymatically transformed into the herbicidally active L-isomer to create a resolved isomer of L-glufosinate ammonium. Liberty ULTRA is an improved version of Liberty 280 with innovations from both chiral chemistry and formulation chemistry. Field trials were conducted from 2021 to 2023 to compare weed control efficacy between Liberty ULTRA and Liberty 280. Liberty ULTRA at 370 g ai/ha demonstrated incremental improvement in overall weed control efficacy compared to Liberty 280 at 654 g ai/ha. Field trials were also conducted in 2022 and 2023 to compare Liberty ULTRA to several generic racemic glufosinate products. Liberty ULTRA at 370 g ai/ha achieved better weed control than all tested generic racemic glufosinate at 654 g ai/ha. Liberty ULTRA herbicide will also feature the Liberty Lock formulation which improves spray droplet retention, increases droplet spreading and ultimately drives more active ingredient into weed leaves compared to generic glufosinate. Liberty ULTRA will have a higher L-glufosinate concentration in the formulation compared to other glufosinate formulations which enables a 25% reduction in application use rate when compared to Liberty 280. The application use rate reduction will mean that customers will be able to make more applications, serve more customers and cover more acres from the same tote or bulk tank compared to most racemic glufosinate herbicides. Liberty ULTRA Herbicide, Powered by Glu-L Technology and the Liberty Lock formulation represents the future of glufosinate for BASF for effective broad spectrum weed control in glufosinate tolerant crops.

STOREN HERBICIDE – PROVIDING MORE CONSISTENCY AND INCREASED LENGTH OF WEED CONTROL IN CORN. Sudeep A. Mathew*[1], Mark Kitt[2]; [1]Syngenta, Germantown, MD, [2]Syngenta, Greensboro, NC (69)

ABSTRACT NOT AVAILABLE

META-ANALYSIS OF MULTIPLE CONSERVATION AGRICULTURAL PRACTICES ON WEED MANAGEMENT IN THE US MIDWEST USING AGEVIDENCE.

Maria A. Gannett*; Cornell University and University of Massachusetts, Ithaca, NY (70)

Conservation agriculture is a set of crop management practices that can conserve environmental resources, first motivated by the extreme erosion experienced during the dust bowl in the United States (U.S.). Reduced tillage was the first practice promoted as a conservation practice, but crop rotation and cover cropping were later included as core conservation agricultural practices. We used a dataset that amassed the results of conservation agriculture research published in the U.S. Midwest from 1980-2020, collected by The Nature Conservancy, to explore the effects conservation agriculture on weed control. The data were standardized by calculating a response ratio comparing more conservative practices to less conservative practices and then analyzed this response using fixed and random effects meta-analysis models. Tillage reduced weed control but cover cropping and crop rotations increased weed control. When all three practices were analyzed together, there was no effect on weed management overall. These results are in line with previous reviews exploring each practice in isolation.

CLIMATE CHANGE IMPACTS ON DAIRY AND FIELD CROP WEED MANAGEMENT.

Caroline Marschner*[1], Sharon N. Bachman[2], Isabella M. Colucci[3], Catalina Ferreira-Dias Rivas[4], Rebecca Stup[1], Claire Liu[5], Xindi Liu[1], Sophie Westbrook[1], Antonio DiTommaso[6]; [1]Cornell University, Ithaca, NY, [2]Cornell Cooperative Extension, East Aurora, NY, [3]Cornell University, Gray, ME, [4]Columbia Mailman School of Public Health, New York, NY, [5]Cornell University, New York, NY, [6]Soil and Crop Sciences Section, School of Integrative Plant Science, Cornell University, Ithaca, NY (71)

Climate change is likely to impact weed management in a variety of ways, from management timing to herbicide efficacy to the competitiveness of weed species. We conducted a series of systematic reviews of literature to find what research is relevant to New York dairy and field crop farmers. The best studied aspect of climate change and weed management is herbicide efficacy. Changes in herbicide efficacy vary depending on mode of action, climate variable studied, and target weed species. There has been extensive research on range shifting species with climate change, but agronomic weeds are underrepresented in the literature for this area. Changes in weed competitiveness are highly variable in the literature, with most species both more and less competitive depending on the climate variable studied and on agronomic system. The clearest trends we observed were for decreased glyphosate efficacy with climate change and increased competitiveness of redroot pigweed, which will impact New York dairy and field crop farmers.

HERBICIDE RESISTANT WEEDS IN NORTH CAROLINA, THEIR DISTRIBUTION, AND MANAGEMENT CONCERNS.

Wesley Everman*[1], Charlie W. Cahoon[1], Eric Jones[2], Jose H. de Sanctis[1], Jackson W. Alsdorf[1], Diego J. Contreras[1], Ronel J. Argueta[1]; [1]North Carolina State University, Raleigh, NC, [2]South Dakota State University, Brookings, SD (72)

[ABSTRACT NOT AVAILABLE]

INTEGRATING FLAME WEEDING FOR EARLY SEASON WEED CONTROL IN ORGANIC SOYBEAN.

Kurt M. Vollmer*[1], Dwayne D. Joseph[2], Alan W. Leslie[3]; [1]University of Maryland, Queenstown, MD, [2]University of Maryland, Chestertown, MD, [3]University of Maryland, College Park, MD (73)

Early season weed control can be difficult in organic production systems. Flame weeding is a non-chemical tactic that has been shown to control several broadleaf weed species. The majority of flame weeding treatments are applied to emerged weeds; however, studies have also shown flame treatments to have detrimental effects on the seeds of certain weed species post-dispersal, including pigweed species. The objectives of this research were to 1) determine the depth of burial and amount of flame exposure (speed of application) needed to kill weed seeds, and 2) evaluate flame-weeding as an integrated tactic for early season weed control. In order to evaluate objective 1, a two-factor factorial study was conducted at the UMD research greenhouse in College Park, MD. The first factor consisted of smooth pigweed seeded in aluminum trays at 0, 1.3, 2.5, and 3.8 cm below the soil surface. The second factor consisted of flaming the trays were at 1.6, 3.2, and 4.8 kph. A non-flamed control was included for comparison. Results showed that smooth pigweed emergence was reduced when seed was buried deeper in the soil profile (2.5 cm to 3.8 cm), but there were no clear patterns related to time of flame exposure. In order to evaluate objective 2, a two factor factorial study was conducted in two organic soybean fields in Kent and Caroline county in Maryland. The first factor consisted of flame and/or cultivation treatments. All plots received a pre-emergence flame treatment followed by no additional treatment, one sequential flame treatment, two sequential flame treatments, a cultivation treatment, or a cultivation treatment followed by an additional flame treatment when weeds reached 7.6 cm height. The second factor evaluated flame exposure at 1.6 and 3.2 kph. At 5 weeks after planting at the Kent County site, pre-flaming followed by 2 sequential flame treatments reduced overall broadleaf weed density (26% of untreated) compared to pre-flaming alone (90% of untreated) and pre-flaming followed by 1 sequential flame treatments (55% of untreated), but pre-flaming followed by cultivation treatments caused the greatest reduction in early season weed density (11% of untreated). There was a significant treatment by flame exposure interaction for pigweed density at 4 weeks after planting the Caroline county site. Pre-flaming followed by cultivation treatments caused the greatest reduction in early season pigweed density (25% of untreated), regardless of flame exposure. Pre-flaming followed by cultivation followed by an additional flame treatment provided a similar reduction in weed density as cultivation only treatments when applied at 1.6 kph (39% of untreated) compared to 3.2 kph (81% of untreated). These results indicate that although flame-weeding has some effect on reducing early season weed density, it does not provide acceptable, long-term weed control in organic soybean. This may be due the variability in weed emergence among individual species. Further research should evaluate timing of flame-weeding treatments for optimum effectiveness on individual species.

AN OVERVIEW OF PYRIMISULFAN PRODUCTS FOR TURF AND ORNAMENTAL WEED CONTROL.

Shawn Askew*[1], Eric Reasor[2], Dale Sanson[3]; [1]Virginia Tech, Blacksburg, VA, [2]PBI Gordon Corporation, Rowlett, TX, [3]PBI Gordon Corporation, Kansas City, MO (74)

Pyrimisulfan is a pyrimidinyl(thio)benzoate herbicide that inhibits acetolactate synthase and is marketed under the trade names Vexis™ and Arkon™ for granular and liquid formulations, respectively. Pyrimisulfan was registered for rice production in Japan in 2010 and PBI Gordon Corporation started research for potential uses in the US turfgrass and ornamentals markets in 2009. A series of formulations improvements led to a variety of clay-based granular carriers and a soluble concentrate upon which the Vexis™ and Arkon™ herbicide products are based. Key pests controlled by pyrimisulfan include yellow nutsedge, purple nutsedge, green and false green kyllinga, and dollarweed. The product suppresses or controls a number of other broadleaf and sedge weeds. Key market fit for pyrimisulfan includes sedge control that is competitive with market leaders, improved residual sedge control, improved performance against sedges from a granular product, registration of the sprayable product on golf greens, excellent safety to both warm- and cool-season turfgrasses, and market-leading potential for dollarweed in the Southern US. Like several other acetolactate synthase inhibitors currently marketed for sedge control, pyrimisulfan use rates are low, not exceeding 70 g ai ha^{-1} and the product can be successfully mixed with a wide range of other products without compromising weed control. Future goals for pyrimisulfan include exploring potential uses in ornamental plant product and landscaping and industrial vegetation management.

APPLICATION STRATEGIES USING ARKON (PYRIMISULFAN) FOR CONTROL OF FALSE-GREEN KYLLINGA AND YELLOW NUTSEDGE.

Brian A. Aynardi*[1], Matthew T. Elmore[2], Steven J. McDonald[3], Eric Reasor[4], Jeffrey W. Marvin[5], Robert C. Williamson[6], Chrissie A. Segars[7]; [1]PBI-Gordon Corporation, State College, PA, [2]Rutgers University, New Brunswick, NJ, [3]Turfgrass Disease Solutions, LLC, Spring City, PA, [4]PBI Gordon Corporation, Rowlett, TX, [5]PBI-Gordon Corporation, Overland Park, KS, [6]PBI-Gordon Corporation, Defiance, OH, [7]PBI-Gordon Corporation, Bend, OR (75)

Pyrimisulfan is an acetolactate synthase (ALS) inhibitor registered and sold by PBI-Gordon Corporation. It is registered for use in a variety of turfgrass settings for control of numerous sedge, kyllinga, and select broadleaf weed species, including: yellow nutsedge (*Cyperus esculentus* L.), purple nutsedge (*Cyperus rotundus* L.), green kyllinga (*Kyllinga brevifolia* Rottb.), false-green kyllinga (*Kyllinga gracillima* Miq.), cockscomb kyllinga (*Kyllinga squamulata* Vahl), and dollarweed (*Hydrocotyle umbellata* L.). In contrast to many of the existing ALS herbicide active ingredients registered in turf, pyrimisulfan is available in both liquid and granular formulations, and may be used on cool and warm-season turfgrass species, as well as on putting greens (liquid only). Application strategies evaluating single treatments at 70 grams ai ha^{-1} and sequential treatments at 52.5 grams ai ha^{-1} have demonstrated unique benefits for turfgrass managers. A significant reduction of tubers in both *C. esculentus* and *C. rotundus* has been observed under controlled conditions. Additionally, pyrimisulfan provided near complete control of rhizome regrowth with *K. gracillima* under controlled conditions. These studies support field efficacy trials where early post-emergent applications of pyrimisulfan provide near season-long control of these problematic weeds.

SPRING APPLIED PINOXADEN TOLERANCE ON CREEPING BENTGRASS, FINE FESCUE, POA ANNUA, PERENNIAL RYEGRASS, TALL FESCUE AND LITTLE BLUESTEM IN THE MID-ATLANTIC.

Steven J. McDonald*; Turfgrass Disease Solutions, LLC, Spring City, PA (76)

The pinoxaden label was recently expanded to include several cool-season turfgrass species. Six field trials were conducted to evaluate single and sequential applications of pinoxaden (applied as Manuscript, Syngenta Crop Protection) to *Agrostis stolonifera* (L-93), *Festuca longifolia* (Beacon), *Lolium perenne* (Evening shade), *Festuca arundinacea* (Bullseye), *Poa annua* and *Andropogon spp.* for tolerance in May of 2023. Rates of pinoxaden tested ranged from 0.0151 kg a.i./ha (4.1 fl oz/A product) to 0.141 kg a.i./ha (38.4 fl oz/A product) and were mixed with Adigor surfactant at 0.5% v/v. All treatments were replicated four times and arranged in a randomized complete block design. Data from the spring studies show that pinoxaden has excellent safety on *P. annua*, *F. longifolia*, and *Andropogon spp.*, with no injury observed at any of the tested rates. The tolerance of *L. perenne*, *F. arundinacea* and *A. stolonifera* depended on rate and application frequency. In general, tolerance decreased as rates increased for all species, however plant safety was observed with lower rates (< 0.023 kg a.i./ha) of pinoxaden on *Agrostis*. This data suggests pinoxaden may be an effective tool for removal of grassy weeds in *A. stolonifera* turf, however evaluating additional rates, intervals, application timings, and cultivars is necessary.

CONTROL OPTIONS FOR MULBERRYWEED [FATOUA VILLOSA (THUNB.) NAKAI].

Jeffrey Derr*[1], Aman Rana[2]; [1]Virginia Tech, Virginia Beach, VA, [2]Virginia Tech HRAREC, Virginia Beach, VA (77)

Mulberryweed [*Fatoua villosa* (Thunb.) Nakai] is a summer annual weed in the Moraceae that resembles a mulberry (Morus spp.) tree seedling. It has prominent hairs on the leaves and stems, perhaps leading to another common name of hairy crabweed. It is widespread in southern Asia and was first identified in Louisiana in 1962. It is thought to have spread across the southern United States in the 1990's, probably through the shipment of nursery stock. It grows upright to about 0.3 to 0.9 m tall (one to three feet tall) with alternate, triangular-shaped leaves that are toothed. It flowers very early (3 leaf stage), with plants 5 cm (2 inches) tall capable of producing viable seed. Flower clusters occurs at the leaf axils and are purple. Seed lack dormancy at shedding, resulting in multiple generations during the growing season, leading to a rapid increase in population. Experiments were conducted to evaluate commonly used granular herbicides for preemergence control of this weed. Herbicides tested were pendimethalin, flumioxazin, indaziflam, oxyfluorfen + oryzalin, isoxaben plus trifluralin, pendimethalin plus dimethenamid, flumioxazin plus prodiamine, isoxaben plus dithiopyr, oxadiazon, and oxyfluorfen plus prodiamine. In the first trial, flumioxazin, oxyfluorfen plus oryzalin, and pendimethalin plus dimethenamid provided the numerically highest percent control (98% or higher) at two months after application. All other treatments except isoxaben plus trifluralin gave 90% or greater control In the second study, flumioxazin, oxyfluorfen plus oryzalin, and pendimethalin plus dimethenamid again provided the numerically highest control (essentially complete control), while the other treatments gave 70% or greater control. The treatments evaluated in the postemergence trial were bentazon, halosulfuron, flumioxazin, indaziflam, sulfentrazone, clopyralid, glufosinate, and glyphosate. A nonionic surfactant (Capsil) was added to all treatments. In the first trial, flumioxazin gave complete control, sulfentrazone and glyphosate gave fair to good control, while the other treatments gave unacceptable control at 4 WAT. In the second trial, flumioxazin, sulfentrazone, glufosinate, and glyphosate all gave excellent control while the other treatments gave unacceptable control. The greater control in the second trial was probably due to a smaller plant size compared to the first trial. There are available control options for this weed. Care must be taken to control escapes from either preemergence or postemergence herbicide application, since surviving plants can quickly produce viable seed, which can then germinate after being shed.

EFFICACY OF FOLIAR APPLICATIONS ON AUTUMN OLIVE.

Jeff Jodon*; Penn State University, University Park, PA (78)

EFFICACY OF FOLIAR APPLICATIONS TO AUTUMN OLIVEJ.C. Jodon Jr.*, J Sellmer, Penn State University, University Park, PA ABSTRACT Autumn olive (*Elaeagnus umbellata Thunb.*) is a spreading and colonizing invasive shrub found along roadsides in Pennsylvania. *Elaeagnus umbellata Thunb.* is a small tree or multi-stem shrub, capable of fixing nitrogen, which aids its establishment and growth in poor soil conditions found along the roadside. An experiment was established at the Penn State Russell E. Larson Agricultural Research Center, Agronomy Farm in Rock Springs, PA. The herbicide treatments included: 3193 g 2,4-D ae/ha + 280 g aminocyclopyrachlor ae/ha + 21 g metsulfuron-methyl ae/ha; 3193 g 2,4-D ae/ha + 1681 g triclopyr ae/ha + 21 g metsulfuron-methyl ae/ha; 280 g aminocyclopyrachlor ae/ha; 1681 g triclopyr ae/ha; 3361 g triclopyr ae/ha; 10,084 g triclopyr ae/ha; 3193 g 2,4-D ae/ha; 4257 g 2,4-D ae/ha; 2241g dicamba ae/ha; 21 g metsulfuron-methyl ae/ha; and an untreated check. The experiment was established as a complete randomized design with ten plants per treatment. Individual shrubs were measured, the average width was multiplied by the height which was then doubled to approximate the entire canopy area of each plant. Herbicide application dosage per plant was based on calculated canopy area. By the last rating, several treatments had missing plants for various reasons, however, trees falling on test plants was the main reason. The final number of replicates for the treatments ranged between 7-10. All herbicide treatments were applied at 327 L/ha and included methylated seed oil at 1% v/v. Treatments were applied using a CO_2-powered backpack sprayer equipped with a 30 GunJet handgun with one PPX 6 nozzle at 35 psi. Plants were treated on September 19, 2019. Treatments were visually rated for percent injury where 0 = no injury– 100 = complete injury on October 3, 2019, 14 days after treatment (DAT), and percent living leaf canopy using a similar rating system occurred on May 5, September 22, 2020, and September 15, 2021, 229 DAT, 370 DAT and 728 DAT, respectively. All data were subject to analysis of variance and when treatment F-tests were significant ($p \leq 0.05$), treatment means were compared using Tukey's HSD separation test. By October 3, 2019, 14 DAT, percent injury ranged from 4% to 99.7%. All herbicide treatments, except for 21 g metsulfuron-methyl ae/ha (4% injury), showed similar levels of injury. Although 1681 g triclopyr ae/ha (99.7%) produced the highest injury 14 DAT; by 229 DAT treatment metsulfuron-methyl at 21 g ae/ha resulted in the lowest living leaf canopy with 0% followed by 2241g dicamba ae/ha at 0.3%. The least effective treatment was 4257 g 2,4-D ae/ha with 22.9% living leaf canopy. Except for 21 g metsulfuron-methyl ae/ha, all other treatments showed signs of resprouting from dormant buds or roots at 229 days after treatment. This trend continued at the one- and two-year ratings after treatment. Overall, treatments resulting in the lowest living leaf canopy was at 21 g metsulfuron-methyl ae/ha (0%) and 280 g aminocyclopyrachlor ae/ha (4.3%). We found that herbicide mixes including 21 g metsulfuron-methyl ae/ha were less effective than 21 g metsulfuron-methyl ae/ha applied alone which suggests that antagonism is occurring within the mixes reducing metsulfuron-methyl effectiveness.

CONTROLLING COMMON RAGWEED IN FRASER FIR WITH WHITE CLOVER LIVING MULCH.

Joseph C. Neal*[1], Christopher D. Harlow[1], Jeff Owen[2], Brad Edwards[3]; [1]North Carolina State University, Raleigh, NC, [2]North Carolina State University, Mills River, NC, [3]North Carolina Cooperative Extension, Jefferson, NC (79)

Controlling common ragweed (*Ambrosia artemisiifolia*) in Southern Appalachian Fraser fir Christmas tree production is complicated by the desire to maintain a living ground cover dominated by white clover. Currently labeled postemergence herbicides are either too injurious to the white clover, too injurious to the Fraser fir trees when ragweed is at optimal growth stages, or are ineffective on ragweed. As part of an ongoing effort to find solutions for this vexing problem, on-farm experiments were conducted to evaluate preemergence (PRE) and postemergence (POST) herbicides for ragweed control in established Fraser fir. All herbicides were applied as semi-directed sprays contacting the lower 1/3 of the trees. Preemergence and postemergence tests were established in separate Christmas tree farms in Laurel Springs, NC. The trees, planted in 2021, were 12 to 24 inches tall. PRE herbicides were trifloxysulfuron, cloransulam, topramezone, flazasulfuron, flumioxazin, fluridone, indaziflam, oxyfluorfen, and diclosulam, each applied at labeled doses. POST herbicides were 2,4-D amine at 1 lb ai/A, 2,4-DB at 1 lb ae/A, cloransulam at 0.3 & 0.75 oz/A, clopyralid at 0.25 lb ai/A, mesotrione at 3 oz/A, and topramezone at 2 & 4 oz/A. Herbicides were applied with a CO_2-pressurized sprayer equipped with two TeeJet 11002 TTI nozzles and calibrated to deliver 15 GPA. Herbicides were applied to the row middles on both sides of a row of Fraser fir trees. Spray overlap in the crop row also contacted the lower ~12 inches of the trees. PRE herbicides were applied on April 4, 2023, before Fraser fir trees had broken bud. POST herbicides were applied on July 17, 2023, at which time the ragweed was mostly 12 to 24 inches tall, and new growth on Fraser fir plants had fully elongated with terminal buds on the branches set. Injury to Fraser fir and ragweed control were visually evaluated. No injury to Fraser fir trees was observed from PRE or POST herbicides. The only PRE herbicides to control ragweed were flumioxazin or diclosulam. Those treatments also caused little or no persistent injury to white clover. Diclosulam also controlled horseweed. In the POST study, mesotrione, 2,4-DB, and topramezone at 2 oz/A did not control ragweed. Grower-acceptable levels of ragweed control were obtained with topramezone at 4 oz/A, cloransulam, clopyralid, or 2,4-D amine. Concerns remain over topramezone and clopyralid safety to clover living ground cover, as well as consistency of ragweed control with cloransulam. Furthermore, Fraser fir developmental stages where plants are not injured by semi-directed applications of 2,4-D amine need further study. In this POST study, the new growth on Fraser fir was more mature than normally observed in mid-July, likely related to higher-than-average temperatures and lower-than-average rainfall in the month preceding herbicide applications. Controlling common ragweed remains a challenge for growers who wish to maintain a living ground cover for erosion control. It appears that sequential applications of preemergence and postemergence herbicides will be required to obtain satisfactory control with adequate safety to the crop and living ground cover.

EIGHT YEARS OF HARVEST WEED SEED CONTROL RESEARCH IN VIRGINIA.

Michael L. Flessner*; Virginia Tech, Blacksburg, VA (80)

Since 2015, studies examining various aspects of harvest weed seed control (HWSC) in Virginia indicate a substantial opportunity to reduce weed seedbank inputs in soybean and wheat production. Weed seed retention, a requirement of HWSC, is high (>80% for many weed species and >90% for certain key species like Palmer amaranth (Amaranthus palmeri S. Watson), combines successfully get weed seeds into the HWSC device (>95% of seeds that enter the combine), and seed impact mills kill >97% of weed seeds that enter (except Italian ryegrass (Lolium perenne L. ssp. multiflorum (Lam.) Husnot) that is >90% kill). Various studies indicate about a 30% reduction in weed density in the season following one HWSC event, although wide variations have been observed. Testing of chaff lining and seed impact mills on commercial farms has had some issues of material plugging the combine when harvesting green or high moisture residues. Class 8 combines have sufficient engine capacity to power seed impact mills.

MANAGING ROLLED-CRIMPED COVER CROPS FOR ORGANIC NO-TILL WHEAT (TRITICUM AESTIVUM).

Uriel D. Menalled*[1], Sandra Wayman[1], Terry J. Rose[2], Christopher Pelzer[1], Matthew Ryan[1]; [1]Cornell University, Ithaca, NY, [2]Southern Cross University, Lismore, Australia (81)

Cover crop mulches can suppress weeds and protect soil from erosion, making them an attractive tool for organic no-till crop production. While cover crop mulches offer numerous benefits for organic no-till systems, they can also lead to yield reductions if not managed appropriately. Choosing cover crop species and management practices that don't reduce the limiting nutrients and the emergence of a cash crop is crucial to the success of organic no-till systems. Five cover crop species [buckwheat (*Fagopyrum esculentum* Moench), radish (*Raphanus sativus* L.), sorghum sudangrass [*Sorghum bicolor* (L.) Moench x *S. sudanense* (Piper) Stapf.], soybean (*Glycine max* (L.) Merr.), and sunn hemp (*Crotalaria juncea*L.)] were planted at three dates to assess the use of cover crop mulches for organic no-till winter wheat (*Triticum aestivum*L.) production. The biomass of the cover crop mulches affected weed suppression but not winter wheat establishment or yield. Instead, winter wheat yield was influenced by the cover crop species used to create the mulches. These results indicate that cover crop mulches can suppress weeds without a yield penalty in no-till winter wheat production.

WEED BIOMASS IN THREE COVER CROP BICULTURES IN THE NORTHEASTERN USA. Huong T. X. Nguyen*[1], Richard G. Smith[2], Helen Boniface[3], K. Ann Bybee-Finley[4], Heather M. Darby[5], Sjoerd W. Duiker[6], Masoud Hashemi[7], Sarah M. Hirsh[8], Ivy Krezinski[5], Ellen B. Mallory[9], Tosh R. Mazzone[6], Steven B. Mirsky[4], Thomas Molloy[10], Arthur Siller[7], Resham Thapa[11], Kate Tully[12], Mark J. VanGessel[13], John M. Wallace[6], Nicholas D. Warren[14], Sandra Wayman[15], Mathew R. Ryan[15]; [1]Cornell University, Ithaca, NY, [2]Natural Resources and Environment, University of New Hampshire, Durham, NH, [3]Soil Health Institute, Morrisville, NC, [4]Sustainable Agricultural Systems Lab, USDA-ARS, Beltsville, MD, [5]Plant and Soil Science Department, University of Vermont, Burlington, VT, [6]Department of Plant Science, Pennsylvania State University, University Park, PA, [7]Stockbridge School of Agriculture, University of Massachusetts-Amherst, Amherst, MA, [8]Extension Somerset County, University of Maryland, Princess Anne, MD, [9]School of Food and Agriculture and Cooperative Extension, University of Maine, Orono, ME, [10]Cooperative Extension, University of Maine, Orono, ME, [11]Department of Agricultural and Environmental Sciences, College of Agriculture, Tennessee State University, Nashville, TN, [12]Plant Science and Landscape Architecture, University of Maryland, College Park, MD, [13]" Department of Plant and Soil Sciences, University of Delaware, Georgetown, DE, [14]Natural Resources and the Environment, University of New Hampshire, Durham, NH, [15]School of Integrative Plant Sciences - Soil & Crop Sciences, Cornell University, Ithaca, NY (82)

Cover crop mixtures are popular with farmers seeking to broaden ecosystem services over single species. Understanding the performance of individual species in mixtures is important for optimizing resource partitioning and biomass production. A completely randomized field experiment was conducted across six locations in the northeastern USA in 2020-2021 that compared three cover crop species grown in monocultures and in bicultures: 1) cereal rye, 2) hairy vetch, and 3) forage rape. Weed biomass was collected to assess the weed pressure and land equivalent ratios (LER), and partial LERs were calculated from crop biomass to assess complementarity. Observations show that weed pressure was about fivefold larger in monocultures compared with that in mixtures. Results show that crop species varied in their competitive interactions, with hairy vetch and forage rape sometimes producing higher biomass when grown in mixtures compared to when they were grown separately, whereas cereal rye produced higher biomass and tended to suppress the growth of the other species when grown in mixtures. Environmental conditions across sites also influenced cover crop performance and competitive interactions, indicating the optimal mixtures may vary by location.

WEED SEED PERSISTENCE WITHIN THE PERENNIAL FORAGE PHASE OF A CROP ROTATION.

Carolyn Lowry*; Pennsylvania State University, University Park, PA (83)

When incorporated into crop rotations, perennial forages can help break up annual weed life cycles and decrease populations of annual weeds in the soil seedbank. After the establishment year of a perennial forage, often new seed rain from annual weeds is greatly reduced, while seeds in the soil seedbank are depleted from predation, decay, and aging. We will present highlights from two distinct projects examining how abiotic and biotic factors influence mortality rates of weed seeds in the soil seedbank during the perennial phase of a crop rotation. First, within an alfalfa (Medicago sativa)-orchardgrass (*Dactylis glomerata*) mixture, we examined how warmer temperature associated with climate change may affect weed persistence in the seedbank of four common weed species: *Amaranthus retroflexus*, *Abutilon theophrasti*, *Ambrosia artemisiifolia*, and *Digitaria sanguinalis*. We found that warmer temperatures increased the mortality rate of *A. retroflexus but* had no effect on mortality rates of the other weed species. This suggests that while warmer temperatures from climate change may decrease persistence of some weed species, it is not universal. Therefore, the length of time optimal weed control is required to draw down weed populations in the soil seedbank are unlikely to significantly change. Second, we examined whether perennial forage species and diversity affected weed seed persistence in the soil seedbank. We compared three forage species: alfalfa, orchardgrass, and forage chicory (*Cichorium intybus*), in all combinations of one species, two species, and the three species mixture. We found that as the percent of alfalfa (a legume) in a mixture increased, the persistence of *Amaranthus powellii* decreased, potentially suggesting that nitrogen levels in the soil may influence rates of seed mortality. Future work will examine whether N levels in the soil are drivers of seed mortality, and whether seed mortality was microbially mediated.

EVALUATING A METAMITRON AND ETHOFUMESATE PRE-MIX FOR WEED CONTROL AND CROP SAFETY IN TABLE BEETS AND CARROTS.

Lynn M. Sosnoskie*; Cornell University, Geneva, NY (84)

Table beets are extremely sensitive to early-season weed competition due to their slow and uneven germination and growth. Herbicides are crucial for suppressing weeds in commercial production although few active ingredients are currently available for the crop. Torero, a pre-mix of metamitron (WSSA 5) and ethofumesate (WSSA 15), is registered for both PRE and POST use in beets, internationally. The inclusion of Torero in North American production systems would add an extra active ingredient (metamitron) to an extremely limited chemical toolbox. On 22 June 2023, a research trial was established in 'Ruby Queen' beets at the Cornell AgriTech vegetable research farm in Geneva, NY, to evaluate the efficacy and safety of Torero used both PRE and POST. All herbicide rates and combinations were developed in consultation with the registrant. PRE treatments included Torero (at rates ranging from 2.15 to 9.35 L/ha of formulated product) applied alone and in combination with Dual Magnum (S-metolachlor, WSSA 15, at 0.75 L/ha of formulated product). A nontreated check and a grower standard of Dual Magnum (at 0.75 L/ha of formulated product) and Nortron (at 2.15 L/ha of formulated product) applied PRE were also included in the study. POST treatments included a tank mix (UpBeet [triflusulfuron-methyl, WSSA 2] plus Stinger [clopyralid, WSSA 4] plus Spin-Aid [phenmedipham, WSSA 5] all at labeled rates) of registered herbicides or else Torero at 3.18 L/ha of formulated product. POST applications were made when weeds (Ambrosia artemisiifolia, Chenopodium album, Polygonum spp., Portulaca oleracea, and a mixture of grasses) reached mean heights of 5 to 10 cm. With respect to weed control, all PRE and POST combinations reduced mean weed cover significantly relative to the nontreated check. At 30 to 40 days after planting (DAP), mean weed cover in the nontreated check was 94% compared to 1 to 8% in the herbicide treated plots. With respect to injury, PRE and POST herbicide combinations, mean crop stunting did not exceed 16%. Torero PRE, applied alone or in combination with Dual Magnum, was not more injurious than Dual Magnum plus Nortron. POST applications of Torero did not significantly impact beet growth and development. By 30 to 40 DAP, the greatest amount of crop stunting occurred in the nontreated check plots where competitive interactions were affecting beet growth and development. At end of season, Torero-treated plots were largely weed free; nontreated checks were almost non-harvestable. Preliminary assessments suggest that Torero can provide effective early season weed control in 'Ruby Queen' table beets under the conditions of our trial, which included 12.7 cm of rainfall on a Honeoye loam between 22 June and 30 July 2023. Torero was also safe on 'Ruby Queen' table beets, exhibiting only minor and transient injury. Torero is also compatible with other production practices in NY; this includes the requirement on the label for deep plowing following crop harvest. Additional trials are investigating efficacy and safety in both carrots and onions. Resulting data will be provided to the IR-4 Project and Adama to support future label considerations.

WEED CONTROL EFFICACY AND COLE CROPS TOLERANCE TO TANK-MIXED COMBINATION OF CHLOROACETAMIDE HERBICIDES AND OXYFLUORFEN IN NEW YORK AND NEW JERSEY.

Lynn M. Sosnoskie[1], Wesley M. Bouchelle[2], Thierry E. Besancon*[2]; [1]Cornell University, Geneva, NY, [2]Rutgers University, Chatsworth, NJ (85)

While herbicides are significant components of cole crop production programs, the limited number of registered products and their narrow spectrums of control can result in significant in-season escapes and require the need for costly hand-weeding. Because of crop injury potential, tank-mixes of *S*-metolachlor, a WSSA Group 15 herbicide, and oxyfluorfen are not advised. There are also recommendations to avoid applying oxyfluorfen in cabbage in the same season as Group 15 products. Field studies were conducted in spring of 2022 in New York and fall of 2022 in New Jersey to evaluate weed control and crop tolerance response to herbicide programs by chloroacetamide and diphenylether mixing in cole-crops. Treatments included an untreated weedy control, oxyfluorfen (18 g ha^{-1}) pre-transplant alone or followed by *S*-metolachlor or acetochlor (105 g ha^{-1}) one-day post-transplant, *S*-metolachlor or acetochlor (105 g ha^{-1}) alone or mixed with oxyfluorfen (18 g ha^{-1}) one-day post-transplant, *S*-metolachlor or acetochlor (105 g ha^{-1}) one-day post-transplant followed by oxyfluorfen (18 g ha^{-1}) fourteen days after transplanting. Treatments were arranged as a four replications split-plot design with herbicide treatments as the main plot and cole-crops (cabbage and broccoli) as the subplot. Weed coverage averaged for untreated cabbage and broccoli showed complete coverage (100%). When chloroacetamide and diphenylether herbicides mixes were applied weed coverage averaged only 4% 28 days after transplanting. Injury averaged 16% for cabbage and 20% for broccoli when oxyfluorfen was applied post-transplant, regardless of tank mixing herbicides. Conversely, injury was less than 4% for cabbage and 7% for broccoli when *S*-metolachlor and acetochlor were applied alone post- transplant. Injury following herbicide applied alone post-transplant was greater with *S*-metolachlor (3%) than acetochlor (0.3%) for cabbage, but was = 0.7% for broccoli, regardless of herbicide. Average individual weight of cabbage (1.4 kg) and broccoli (270g) was significantly greater than that for the untreated control (630 g and 130 g for cabbage and broccoli respectively) regardless of the herbicides applied.

IR-4 WEED SCIENCE UPDATE - FOOD CROPS.

Roger B. Batts*; IR-4 Project HQ, NC State University, Raleigh, NC (86)

IR-4 Weed Science Update – Food Crops. Roger B. Batts. IR-4 Project, NC State University, Raleigh, NC Residue projects IR-4 data submitted to EPA led to over 340 new specialty crop uses in 2023, including new or expanded uses for rimsulfuron, trinexapac, ethalfluralin, fluazifop, fomesafen, and acifluorfen. Rimsulfuron uses were approved for pomegranate and edible peel tropical and subtropical small fruits. Trinexapac-ethyl now has established tolerances for both forage and hay clover. Fluazifop approvals include low growing berries, leafy brassica greens, chives, citrus, stone fruits, leaf petiole vegetables, green onions, papaya, head and stem brassicas, and arugula. Fomesafen uses include bulb vegetables, cucurbit vegetables and fruiting vegetables. Glufosinate approvals include grass forage and hay as well as medium to large tropical and subtropical fruits, both edible and inedible peel types. Edamame and low growing berries now have established tolerance for acifluorfen. IR-4 submitted two herbicide data petitions to EPA in 2023 (pyridate and dimethenamid). These submissions could potentially lead to more than 50 new published uses. Twenty-one new herbicide magnitude-of-residue studies began in 2023. They include clethodim on avocado, olive and rice, Ethephon on hazelnut, fluazifop on succulent pea and summer squash, fluroxypyr on mint, glufosinate on peanut and strawberry, indaziflam on asparagus, pyridate on sweet corn, rimsulfuron on avocado, s-metolachlor on perennial peanut, tiafenacil on blueberry, cucumber, mint, pepper and tomato, and uniconazole on greenhouse basil and mint. Nineteen new herbicide and PGR residue studies are scheduled to begin in 2024. These include clethodim on fig, Ethephon on ginseng, ethofumesate on swiss chard, flumioxazin + pyroxasulfone on cucumber, squash and cantaloupe, indaziflam on camas, linuron on mint, stevia and green onion, mesotrione on sesame, metribuzin on potato, NAA on plum and hazelnut, s-metolachlor on carinata and field pennycress, and tolpyralate on hazelnut, blueberry and sweetpotato. Product Performance projects Generating Product Performance (efficacy and crop safety) data to support registration of pest management tools in specialty crops continues to be an important and expanding part of the IR-4 annual research plan. The number of on-going herbicide Product Performance studies in 2023 was thirty-four, including nearly 100 individual trials. The 2024 Performance research plan for herbicides and plant growth regulators includes twenty-seven continuing or new studies (nearly 70 individual trials). Integrated Solutions projects IR-4's Integrated Solutions (IS) Program is structured to assist specialty crop growers outside of the traditional single product/single crop residue and product performance research. IS research efforts focus on crop-pest combinations to address solutions in these four areas, 1) pest problems without solutions, 2) resistance management, 3) products for organic production and 4) pesticide residue mitigation. In 2023, there were seven active IS projects with herbicides and plant growth regulators (18 individual trials), three of which will continue in 2024. Four new weed control IS studies will begin in 2024 (9-10 individual trials), including agave, camelina, pumpkin and hemp.

EVALUATING QUINCLORAC FOR USE IN GRAPES IN NEW YORK AND NEW JERSEY.
Lynn M. Sosnoskie*[1], Thierry E. Besancon[2]; [1]Cornell University, Geneva, NY, [2]Rutgers University, Chatsworth, NJ (87)

Quinclorac, a WSSA Group 4 herbicide, has been registered in other perennial cropping systems for control of field bindweed (Convolvulus arvensis). Group 4 herbicides can cause severe damage to grapes, even at low exposure doses; the response of grapes to quinclorac has not been extensively explored. In 2021 and 2022, IR-4 Project-sponsored research trials were conducted by Cornell University (NY) and Rutgers University (NJ) to evaluate the injurious effects of sequential applications of quinclorac on grape. The NY trial was conducted in Seyval Blanc grapes on 3309 rootstocks planted in 2008 on a Chenango gravel loam (3.3% OM and soil 6.2). The NJ trial was conducted in Cayuga White grapes planted in 2012 on a Woodstown sandy loam (1.63% OM and pH 5.6). Herbicide treatments included applications of quinclorac at 0.43 and 0.84 kg ai ha-1 applied POST two and three times per season. For treatments receiving two applications, herbicides were sprayed 30 days apart with the first application (A timing) approximately 90 days before anticipated commercial harvest. For treatments receiving three herbicide applications, an additional application (C timing) was made after crop harvest/dormancy. All quinclorac applications were banded to the vineyard floor at the base of the vines (on both sides of the vine row), avoiding sensitive bark and foliage. In NY, quinclorac applications caused crop injury (leaf strapping and stunted canopy development) in both 2021 (max. injury 20 to 24%) and 2022 (max. injury 5 to 25%); symptoms persisted throughout much of the growing season. Significant rainfall was recorded in July 2021 (25.9 cm) following the B timing application. On gravelly soil, this may have moved herbicide into the root zone, contributing to injury development. In 2022, observed injury was significantly higher in the 0.84 kg ai ha-1 treatment applied at the A, B, and C timings. Despite observed injury, grape yields were not affected by herbicide treatment. In NJ, minor chlorosis and leaf cupping (<10%) were observed in responses to quinclorac treatments on lower suckering shoots; the damage remained visible until the crop reached dormancy in November. No injury was observed in the grape canopy. Similar results were observed in 2022. Grape yields were not impacted by herbicide treatment. In spring 2023, cordon dieback and stunting of new spurs was noted in NJ, especially at the 0.84 kg ai ha-1 rate averaging 21% and 30%, respectively. In light of these results, future studies are warranted to elucidate how the interactive effects of soil type, rootstock, and rooting depth may affect grape response to quinclorac.

WEED CONTROL AND CROP RESONSE WITH POST-FLOODING DORMANT APPLICATION OF PENDIMETHALIN, FLURIDONE, AND SULFENTRAZONE IN CRANBERRY (VACCINIUM MACROCAPON).

Thierry E. Besancon*, Wesley M. Bouchelle; Rutgers University, Chatsworth, NJ (88)

Pendimethalin, fluridone and sulfentrazone recently received registration for use in cranberry. Previous research indicated that sulfentrazone applied at 420 g ai ha^{-1} at the cabbage head stage increased the rate of terminal bud necrosis. In 2017, the use of pendimethalin in highbush blueberry plantations in New Jersey was associated with widespread phytotoxicity. High pendimethalin use rate, crop shallow root system, edaphic conditions, and herbicide applications timed after blueberry went out of dormancy increased the risk of related injury. Multistate evaluations of pendimethalin on cranberry in 2022 showed that pendimethalin applied at rate > 320 g ha^{-1} and/or after cranberry buds break dormancy could induce severe stunting and reduce berry yield. Fluridone has not been previously investigated for use on cranberry in New Jersey, warranting an evaluation of crop tolerance and weed control efficacy of these herbicides when applied before cranberry breaks dormancy. Treatments were boom-applied at 280 L ha^{-1} at the tight bud stage (April 24) on a 1-yr old nonproductive "Haines" bed in Southampton, NJ, and a 20-yr old productive "Ben Lear" bed in Chatsworth, NJ. Herbicide treatments included pendimethalin at 0.8 or 1.6 kg ha^{-1}, fluridone at 0.22 or 0.45 kg ha^{-1}, and sulfentrazone at 0.28 or 0.42 kg ha^{-1}. For comparison purpose a nontreated control as well as a standard application of napropamide at 6.7 kg ha^{-1} were also included. In the Southampton nonproductive bed, fluridone provided = 97% large crabgrass [*Digitaria sanguinalis* (L.) Scop.] control throughout the season at both rates while control with sulfentrazone decreased to < 50%. By the end of the season (150 DAT), only fluridone at 0.22 or 0.45 kg ha^{-1}, and pendimethalin at 1.6 kg ha^{-1} provided greater control than the napropamide standard. Stunting of new cranberry shoots exceeding 13% was noted 28 and 60 DAT with pendimethalin at 1.6 kg ha^{-1} whereas fluridone or sulfentrazone did not cause significantly greater stunting than the non-treated control or the napropamide standard at both dates. In the Chatsworth productive bed, fluridone application was associated with transient chlorosis (up to 6%) of new cranberry shoots that completely recovered by 60 DAT. As for the other location, pendimethalin application at 0.8 or 4.6 kg ha^{-1} was associated with severe stunting of new growth that culminated 42 DAT with 22% and 36%, respectively, compared to = 7% for other herbicides. With the exception of napropamide that provided 56% control 60 DAT, none of the other herbicides provided control of Carolina redroot [*Lachnanthes caroliana* (Lam.) Dandy], a perennial troublesome of new Jersey cranberry beds. Berry yield was not affected by preemergence herbicides and ranged from 1.6 to 2.4 kg m^{-2}. Overall, fluridone provided better weed control and similar crop tolerance than the napropamide standard.

USE OF SULFENTRAZONE ON NEW CRANBERRY PLANTS IN FIELD AND GREENHOUSE ENVIRONMENTS. Katherine Ghantous*[1], Hilary A. Sandler[2]; [1]UMass Amherst, UMass Cranberry Station, East Wareham, MA, [2]UMass Cranberry Station, East Wareham, MA (89)

Cultivated cranberry (*Vaccinium macrocarpon*) is a long-lived woody perennial with non-deciduous leaves. Preemergence herbicides are applied over the top of dormant cranberry vines and the crop foliage and buds are present during application. The use of sulfentrazone for weed control in cranberry production is relatively new, beginning in 2020. Sulfentrazone carries risk of crop injury on established beds if applications are made when vines are not fully dormant or to beds experiencing high levels of stress from other factors. For other permanent crops, labels stipulate sulfentrazone should only be applied to crops that have been established for three years. The crop safety for applications to young cranberry plantings was unknown. Young beds, characterized by uncolonized, open areas, are particularly vulnerable to weeds becoming established and the use of preemergence herbicides on new beds is an important practice. A greenhouse study was conducted in 2022 to evaluate applications of sulfentrazone on new cranberry plants. Unrooted cuttings from dormant 'Stevens' runners and dormant 1-yr old rooted cuttings (plugs) were used for the study to simulate beds planted with disced in vines or planted with plugs. One of three rates of sulfentrazone (0, 280 g a.i. ha^{-1}, or 420 g a.i. ha^{-1}) was applied at one of two timings (immediately after planting (0 DAP) or 1 week after planting (7 DAP)). Each treatment unit consisted of four unrooted cuttings or four plugs planted into a pot with sand. All treatments were replicated four times, and the study was run twice. Treatments were applied by CO_2powered backpack sprayer. Herbicides were delivered in the equivalent of 3,740 L water ha^1 to simulate application by chemigation. Three months after initial treatments, entire cranberry plants were harvested. For each of the plants grown from unrooted cuttings, the aboveground portion of the plant (shoot) was clipped at the soil level and placed into an individual paper bag. The belowground portion of the plant (root) was gently washed in cool water to remove all of the sand from the roots, and then placed into an individual paper bag. For the plugs, the aboveground portion of the plant was clipped at the at the soil level and placed into an individual paper bag. The roots for the plugs were not quantified due to being unable to remove the original peat/perlite potting media. Visual observations were made for plug roots. Plants were dried in an oven at 60°C for 3 days, then weighed. For unrooted cuttings, the effect of treatment was highly significant ($p = 0.01$), with both root and shoot biomass in treated plants lower than untreated controls for all treatments. Cranberry plants established from rooted plugs tolerated an application of sulfentrazone without significant impacts to growth or widespread visual symptoms of injury, regardless of rate or timing. Aboveground biomass for treated plants was not lower than that of the untreated control. Because the plugs were grown in a peat/perlite media before they were planted into sand for this study, we were unable to quantify root biomass but visual evaluation of roots did not show any differences between treated and untreated plants. In 2023, field trials on a young cranberry bed (planted with rooted plugs in 2021) evaluated three rates of sulfentrazone (0, 280 g a.i. ha^{-1}, or 420 g a.i. ha^{-1}) applied to three different varieties of dormant cranberry. Treatments were applied by CO_2powered backpack sprayer. Herbicides were delivered in the equivalent of 3,740 L water ha^1 to simulate application by chemigation and each treatment was replicated three times per variety. No treatment resulted in any visual symptoms of injury or reduction in fruit yield. Based on these studies, applications of sulfentrazone are not safe for use on new cranberry beds established with unrooted cuttings.

PREEMERGENCE HERBICIDE APPLICATION TIMING ON ROOTING CUTTINGS OF NURSERY CROPS.

Anthony L. Witcher*[1], Isha Poudel[2]; [1]Tennessee State University, Mcminnville, TN, [2]University of Tennessee, Jackson, TN (90)

Container-grown nursery crops are produced from propagative material derived from seedlings or rooted cuttings. Weed issues can originate during propagation and weed seeds are transferred to the subsequent crop. The small container sizes used in propagation increases competition between crops and weeds for light, nutrients, and space which has a negative impact on crop growth and quality. Hand weeding is labor intensive and costly, therefore more cost-effective weed control practices should be adopted. Certain non-root-inhibiting preemergence herbicides have been found safe for use in cutting propagation of certain crop species, but the timing of application may impact crop response. The objective of this study was to determine the effect of preemergence herbicide and timing of application on rooting stem cuttings. In June 2020, containers (16.1 cm diameter square) were filled with a 100% pine bark substrate. Preemergence herbicides (sprayable formulation – isoxaben; granular formulation - isoxaben + dithiopyr and oxyfluorfen + oxadiazon) were applied 2 weeks before sticking stem cuttings (Before Sticking), the day of sticking cuttings (At Sticking), or 2 weeks after sticking cuttings (After Sticking). Crop species included butterfly bush (*Buddleja davidii* 'Nanho Blue'), crape myrtle (*Lagerstroemia indica* 'Catawba'), and holly (*Ilex cornuta* 'Dwarf Burford'). One cutting was stuck in each container, 20 cuttings per treatment, and containers were completely randomized within species and maintained under intermittent mist. Butterfly bush rooting percentage ranged 0% (isoxaben - At and After Sticking) to 75% (oxyfluorfen + oxadiazon - After Sticking) and was 45% for the non-treated control. Rooting percentage was lower for all treatments that contained isoxaben except isoxaben + dithiopyr (After Sticking). Butterfly bush root and shoot dry weight was similar to the non-treated control for all treatments that had rooting. For crape myrtle and holly, no differences were observed (among all treatments compared to the non-treated control) for rooting percentage, root dry weight, total root length, or total root volume. Application timing did not have a significant effect on rooting, thus making applications prior to sticking cuttings would be the most practical and time efficient method for growers. Isoxaben is not labeled for use on butterfly bush, which likely contributed to the damage during rooting. Oxyfluorfen + oxadiazon was safe on all tested crop species, supporting previous research on other crop species during propagation. Isoxaben and isoxaben + dithiopyr were safe on crape myrtle and holly, but all three products should be tested on individual plant species to verify crop safety prior to widespread use in cutting propagation.

PRE-EMERGENCE HERBICIDE PROGRAMS FOR JAPANESE STILTGRASS (MICROSTEGIUM VIMINEUM) CONTROL.

Daniel P. Tuck*, Matthew T. Elmore; Rutgers University, New Brunswick, NJ (91)

Field experiments were conducted in 2022 and 2023 to evaluate various herbicide programs for preemergence Japanese stiltgrass (*Microstegium vimineum*) control in cool-season turfgrass. Research was conducted at the Rutgers Snyder Research Farm in Pittstown, NJ, on a simulated low maintenance lawn composed of tall fescue (*Lolium arundinaceum*) with a natural infestation of Japanese stiltgrass. Treatments were arranged as a two-way factorial design of three herbicides and four application programs. Herbicides consisted of dithiopyr (Dimension 2EW; 280 and 560 g ha^{-1}), prodiamine (Barricade 4FL; 365 and 840 g ha^{-1}), and pendimethalin (Pendulum AquaCap; 1680 and 3360 g ha^{-1}) applied as: 1) a single low rate in March; 2) a single high rate in March; 3) a single high rate in May; 4) and a split application program of a low rate in March followed by a low rate in May. The March applications were made approximately two weeks prior to initial stiltgrass emergence. The May applications were made when stiltgrass was at the two to four leaf growth stage and were designed to coincide with end-user preemergence applications for crabgrass (*Digitaria* spp.) control. Plots measured 0.9 by 2.0 m and were arranged in a randomized complete block design with four replications. March applications were made on 11 March 2022 and 17 March 2023 and May applications were made on 11 May 2022 and 19 May 2023. Treatments were applied using a CO_2-powered single AI9504EVS nozzle boom with 410 L ha^{-1} of water carrier. Japanese stiltgrass control was evaluated visually on a 0 (i.e., no control) to 100 (i.e., complete control) percent scale relative to a nontreated control. Data were analyzed using the GLIMMIX procedure in SAS ($P = 0.05$) and Fisher's protected LSD test was used to separate means. Data were combined across years as no main effect by year interactions were detected. Both main effects of herbicide and application program were significant, however, a main effect interaction was not detected. Averaged across application timings, pendimethalin programs provided greater Japanese stiltgrass control (93%) than prodiamine programs (87%), and dithiopyr programs provided 90% control. Averaged across herbicides, single high rate March herbicide applications and split herbicide application programs provided greater Japanese stiltgrass control (95%) than single low rate March herbicide applications and single high rate May herbicide applications (83 and 87%, respectively). This research indicates preemergence herbicides commonly used for crabgrass control will provide acceptable Japanese stiltgrass control if the application is made before emergence. Future research should investigate early postemergence control with preemergence herbicides.

ANNUAL BLUEGRASS ENCROACHMENT INTO COOL-SEASON TURFGRASS TRIALS SUBJECTED TO TRAFFIC.

Bradley S. Park*, James A. Murphy; Rutgers University, New Brunswick, NJ (92)

Annual bluegrass (ABG; *Poa annua* L.) can invade turfgrass stands thinned by traffic stress. Annual bluegrass encroachment into highly trafficked turfgrass sites can be potentially mitigated by establishing turfgrass cultivars with improved traffic tolerance. The objective of our field trials was to evaluate ABG cover in turfgrass cultivars and experimental selections in the National Turfgrass Evaluation Program (NTEP) Kentucky bluegrass (*Poa pratensis* L.) and tall fescue (*Lolium arundinaceum* [Schreb.] Darbysh.) trials subjected to traffic stress in North Brunswick, NJ. Traffic was applied as a strip across half of each plot in both trials using a combination of the Rutgers Wear Simulator and the Cady Traffic Simulator; the other half of each plot did not receive traffic. Annual bluegrass cover (0 to 100% scale) and uniformity of turf cover (1 to 9 scale; 9 = most complete, uniform cover) were visually evaluated on Kentucky bluegrass and tall fescue entries after each trial received 84 traffic machine passes during 2022. Annual bluegrass cover and uniformity of turf cover data were analyzed as 2 (no traffic and traffic) x 89 (entries) and 2 x 132 factorial strip-plot designs for the Kentucky bluegrass and tall fescue tests, respectively. Greater ABG cover was observed in traffic plots compared to no traffic plots for both species trials; non-trafficked plots had better uniformity of turf cover compared to trafficked plots for both species trials. ANOVA detected a significant traffic × entry interaction for ABG cover and uniformity of turf cover in the Kentucky bluegrass trial and a significant entry main effect for both parameters in the tall fescue trial. Kentucky bluegrass entries with the least (2 to 13%) ABG cover under the trafficked condition were Barvette HGT, Selway, PST-K15-167, DLFPS-340/3500, DLFPS-340/3549, J-1138, Blue Knight, NuRush, PST-K15-172, Yellowstone, Kenblue, RAD 553, BAR PP 79366 and DLFPS-340/3494; Barvette HGT and DLFPS-340/3500 had the best uniformity of turf cover under the trafficked condition. One-hundred-eight (108) tall fescue entries had the least (15 to 28%) ABG cover; cultivars and experimental selections the least ABG cover and the best uniformity of turf cover were GLX ACED, Finelawn Supreme, PST-5GQ, Spyder 2LS, PST-5TRN, Gallardo, Monument, TD2, PPG-TF-267, PPG-TF-249, Lifeguard, Topshelf, Essential 2, Fayette, DLFPS-321/3703, Bullseye LTZ, Endgame, Rover, Tank, ATF 1768, PPG-TF-231, Stealth, Falcon Supreme, 3B2, Padre 2, Firenza II, Providence, K18-RS6, Hemi, Fairfield, AST8118LM, Raptor LS, Extravaganza, PST-5THM, JT 233, ProGold, Bullseye, Serenade, O'Keefe, Birmingham, Firecracker G-LS, BAR FA 8228, Titan GLX, RDC, Hellcat GLR, RAD-TF0.0, Escalade, Bravo 2, Naturally Green, Xanadu, Titan MAX, SETFM2, PPG-TF-337, Expanse, Daybreak, Kizzle, PST-5DC24, SETFM3, Paramount, Grand Prix, Raptor III, Moondance GLX, A-TF31, 5LSS, Palomar, Oriole, Gro-Pro, and Battle Hawk. Annual bluegrass encroachment and uniformity of turf cover are among the criteria that can be used to select Kentucky bluegrass and tall fescue cultivars for high traffic sites; other important criteria include turfgrass quality, spring green-up, genetic color, disease susceptibility, and type classification (for Kentucky bluegrass).

BENTGRASS ENCROACHMENT INTO AN ANNUAL BLUEGRASS FAIRWAY VARIES BY CULTIVAR.

Matthew T. Elmore*, Daniel P. Tuck; Rutgers University, New Brunswick, NJ (93)

A field experiment was conducted to determine the competitiveness of different bentgrass (*Agrostis* spp.) cultivars established vegetatively into an annual bluegrass (*Poa annua*) fairway. This experiment was conducted at the Rutgers Horticulture Farm No. 2 in North Brunswick, NJ from 2018 to 2023. Treatments consisted of eight bentgrass cultivars. Seven creeping bentgrass (*Agrostis stolonifera* L.) cultivars were selected to represent a gradient from of low density, older cultivars to higher density, modern cultivars and consisted of 'Penncross', 'L-93', '007', 'Shark', 'Luminary', 'Pirhana', and 'Flagstick'. 'Puritan' colonial bentgrass (*Agrostis capillaris* L.) was also included. The site was an annual bluegrass fairway maintained at a 95 mm mowing height. The site was irrigated, treated with fungicides, insecticides, and wetting agents to optimize annual bluegrass growth. Nitrogen was applied foliarly at 5 to 10 kg ha^{-1} from May through September as needed to maintain fullness of turf cover. We characterize the annual bluegrass on this site as a weak perennial. It survives the summer under careful management and exhibits characteristics of var. *reptans* [*Poa annua* var. *reptans* (Hauskins) Timm.]. Treatments were arranged in an RCBD with four replications with plots sized 1.0 by 1.0 m. Bentgrass was established to each plot by cutting four 8 by 90 cm strips (2.5 cm depth) of a particular cultivar from a bentgrass nursery and installing them in the annual bluegrass fairway in parallel, spaced 15 cm apart. Annual bluegrass and creeping bentgrass cover were evaluated using two different methods of grid intersect count. One method used a grid that contains 40 intersects each directly over the centerline of the four installed bentgrass strips and 30 intersects each at 2.5, 5, and 7.5 cm from the edge of the strips to evaluate percent bentgrass cover at various distances from the original vegetative planting. The other method, used only in November 2022 and May 2023 used a grid with 100 intersects evenly spaced within a 0.9 by 0.9 m square. Data were analyzed using the GLIMMIX procedure in SAS (P = 0.05) and Fisher's protected LSD test was used to separate means. Turfgrass cover was 100% in all plots on all rating dates, thus, cover reported as not bentgrass was annual bluegrass. Across all evaluations from Jun 2021 until May 2023, Shark and Pirhana provided the most bentgrass cover. On most dates, bentgrass cover provided by 007 and Luminary was similar to Pirhana and Shark. In November 2022, Shark, Pirhana, and 007 were in the top statistical category with 88 to 94% bentgrass cover. Penncross provided less creeping bentgrass cover than these cultivars (74%). By May 2023, Luminary joined Shark, Pirhana, and 007 in the top statistical category with these four cultivars providing 84 to 92% bentgrass cover. Penncross and L-93 provided less bentgrass cover (70 and 78%, respectively) than Flagstick, Luminary, Shark, Pirhana, and 007 in May 2023. Flagstick provided less bentgrass cover than Shark and Pirhana but more than Penncross and L-93 in May 2023. Puritan colonial bentgrass generally provided cover similar to Penncross throughout the experiment. Newer bentgrass cultivars with improved density are more competitive against annual bluegrass in a fairway than older cultivars such as Penncross and L-93.

DISSIPATION OF SPRING-APPLIED METHIOZOLIN IN TURFGRASS SYSTEMS.

Shawn Askew*[1], John M. Peppers[1], Suk Jin Koo[2], Ki-Hwan Hwang[3]; [1]Virginia Tech, Blacksburg, VA, [2]Moghu Research Center, Daejeon, South Korea, [3]Moghu Research Center Ltd., Daejeon, South Korea (94)

Methiozolin is applied five or more times per year to control *Poa annua* in cool, temperate areas, but high market demand in the Southern US and recent registration in Australia expands the products use in variable climates. To better design weed control programs for variable turf types, more information is needed to characterize methiozolin dissipation in different turf systems. Methiozolin was applied biweekly thrice to a *Poa pratensis* lawn and adjacent bare soil in New Jersey and on twelve *Cynodon transvaalensis* x *dactylon* putting greens in Virginia. Soil samples were collected immediately following each application and biweekly for 12 additional weeks. Methiozolin was extracted from each soil sample and analyzed using liquid chromatography with tandem mass spectrometry. Methiozolin was detected only within the top 2 cm of the soil (including verdure), but not below 2 cm, demonstrating limited vertical mobility. Dissipation was significantly faster in turf-covered soil compared to bare soil. The time required for 50% methiozolin dissipation was 13 and 3.5 d in bare soil and turf-covered soil, respectively. In Virginia, methiozolin dissipation in the one-month span of three sequential applications differed between years. Methiozolin concentration immediately following the third biweekly application to *C. transvaalensis* x *dactylon* greens was approximately 105 and 180% of the concentration immediately following the initial application, in 2021 and 2022, respectively. This difference in methiozolin accumulation following three applications was attributed to differential *C. transvaalensis* x *dactylon* greenup during methiozolin treatments each year. Despite differences in post-treatment methiozolin concentration between years, the temporal dissipation rate later into the summer was consistent. Following the final application on *C. transvaalensis* x *dactylon* greens, methiozolin dissipated 50 and 90% in 14 and 46 d, respectively. These data suggest that methiozolin dissipates more rapidly in turfgrass systems than in bare soil.